T0093337

The Scientific Truth, the Whole Truth and Nothing But the Truth

There is a limited understanding amongst scientists, students, and the public about realizing trust in scientific findings. This should be a paramount objective. Scientists and the public need to know more about the link between the philosophy of science and science research methods. There is a limited understanding of why accuracy is important and that it is not the same as precision. Also, there is often the need to be pragmatic and so measure an approximation of a real system, and the classic case is reductionism in biology versus whole organism biology. The author brings these topics together in terms of trusting in science.

Features

- Covers how scientific truth is perceived and increases the preparedness of early career scientists.
- Examines the relatively new field of machine learning and artificial intelligence as applied to crystallography databases in biology and chemistry for new discoveries.
- Describes the major changes in digital data archiving and how vast "raw data" archives are being increasingly developed for machine learning and artificial intelligence as well as complete truth.
- This unique volume will be of interest to pre-university and university undergraduate students, principally in science.
- Presents scientific research examples from physics, chemistry, and biology together with their methodologies.

The Scientific Truth, the Whole Truth and Nothing But the Truth

John R. Helliwell
Emeritus Professor of Chemistry University of Manchester

CRC Press is an imprint of the
Taylor & Francis Group, an **informa** business

First edition published 2024
by CRC Press
2385 Executive Center Drive, Suite 320, Boca Raton, FL 33431

and by CRC Press
4 Park Square, Milton Park, Abingdon, Oxon, OX14 4RN

CRC Press is an imprint of Taylor & Francis Group, LLC

ISBN: 9781032521398 (hbk)
ISBN: 9781032521077 (pbk)
ISBN: 9781003405399 (ebk)

DOI: 10.1201/ 9781003405399

Typeset in Minion
by Deanta Global Publishing Services, Chennai, India

Frontispiece, let me explain:-

Society expects the truth, the whole truth and nothing but the truth.

And, two Quotations

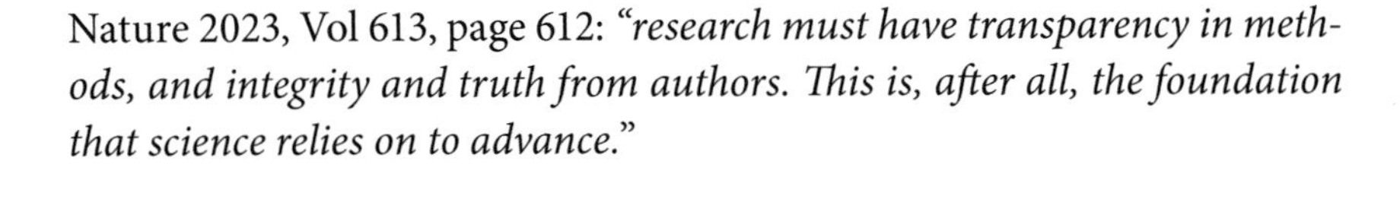

Nature 2023, Vol 613, page 612: *"research must have transparency in methods, and integrity and truth from authors. This is, after all, the foundation that science relies on to advance."*

****

"Science gave me a route to a more rational understanding of the world. It gave me greater certainty too, stability even, and a better way to pursue truth; the ultimate objective of science."

Paul Nurse (2020) page 50 in the ebook "*What is life? Understanding Biology in Five Steps*" Print length 131 pages. Published by David Fickling Books, Oxford, UK.

Contents

Acknowledgements

SEVERAL PEOPLE HAVE HELPED me by commenting on drafts of this book. Firstly, to Brian McMahon, who also made comments on my book proposal. To Hilary Lafoe, who was encouraging and marshalled the referees of my book proposal into providing thoughtful and yet constructive reports on it. To James Kaduk, who came and asked, "can I join you for breakfast" at a Starbucks near the Baltimore Waterfront Hotel during the American Crystallographic Association's 2023 annual conference. This led to my learning about his studies of the psilocybin polymorphs and the courtroom dispute about them. To Kat Bazeley, MRVS (Member of the Royal Veterinary Society), my sister-in-law, who has made helpful comments on the near-final draft. To Sukirti Singh of Taylor & Francis, India, for her patience in smoothing off the rough edges on my submitted drafts. I am grateful to Bharath Selvamani and colleagues for their care in the proofs stages of my book. I thank Ann Chapman for coordinating the production of my book.

I dedicate a heartfelt thanks to all my students and research staff, in the universities and in the scientific civil service, and to the many colleagues I have met in conferences and in my representational and community service roles, who have all greatly enriched my interests and my understanding, which I hope shine through my book as practical experience. Any errors or misconceptions are, of course, my own.

The references are organized chapter by chapter in the book. I provide subject and name indexes, which I think will be useful.

Preface

WHY IS MY BOOK needed? There is no book that spans all the following content together:

I. There is a limited understanding amongst the public, students, and even scientists themselves about realizing trust in scientific findings. The philosophy of science can help.

II. Having worked in the UK's scientific civil service and academic departments of physics and chemistry, I have developed a broad range of scientific interests including biology. On this basis, I provide an atlas of the science subjects.

III. There is a limited understanding of why accuracy is important, and why it is not the same as precision.

IV. I provide case studies focusing on the probes of the structure of matter and from the biological, chemical, materials, and physical sciences. A particularly interesting case is where scientific truth meets legal truth, namely in the courtroom, and I provide a pharmaceuticals dispute case study highlighting this.

V. There are major changes in digital data archiving, especially capacity, thus allowing big (raw) data analyses. In addition, for much longer, vast databases have emerged, such as the Cambridge Crystallographic Data Centre with over one million crystal structures and the Protein Data Bank with more than 200,000 depositions. These are being increasingly harnessed for machine learning and artificial intelligence.

VI. I have undertaken a considerable amount of peer review work, including as Editor in Chief of the International Union of Crystallography's (IUCr) Journals, Main Editor of Crystallography Reviews published by Taylor & Francis, and Chairman of the IUCr OUP Book Series

Committee. This is the role of gatekeeper of what, on publication, are the versions of record. Editors are a pillar of how science establishes truth in research findings.

VII. Notwithstanding this last point pre-publication peer review can fail and post publication peer review is invoked. A rather rare occurrence in my experience, it is nevertheless a methodology that is becoming institutionalized by the preprint server and journals devoted to immediate publication and then open peer review.

VIII. As the scope of data archiving has expanded, it has become possible to consider archiving all data. Thus, the possible worry over what portion of the data has been selected and why can be avoided.

IX. My conclusions and outlook for the whole book will bring these topics together along with the views and activities of leading scientific organizations, such as the USA National Academies of Science, Engineering, and Medicine and the Nobel Foundation including their worries about misinformation and disinformation of science as demonstrated in their excellent webinars and workshops on these topics.

X. In this part, I describe how my career as a retired, or, may I say, semi-retired, scientist fits with my continuing efforts with science, such as in my peer review. When asked, I give my permission for my referee reports to be open and with my name although this latter, even though I have given my permission, is not always taken up by the journal with no reason given.

Obviously, the title of my book takes advantage of my series of books about the scientific life thereby extending my series of topics.

So, as a one-line, non-technical description of my book using layman's terms, I would describe it as follows: *I explain when, in science, are its findings true and its results properly reported.*

What is my book not about?

It is not a comprehensive treatise about scientists deliberately falsifying experiments or analyses i.e., making untruths in seeking personal gain or glory. That said, I cannot ignore it as a topic altogether. So, I will describe where chemical crystallography did get caught out in 2007 when falsified data files were made to make a new publication. There is also a middle ground, which is how to deal with disinformation by those trying to manipulate public opinion. Misinformation is the more innocent mode of sharing, albeit careless, than deliberate disinformation. I will describe the important USA National Academies of Science, Engineering, and Medicine and Nobel Prize Foundation webinar on Truth, Trust, and Hope (2023).

Why do I think this book will matter? As an educator, of course, I want my book to be useful to early-career science researchers and pre-university and university undergraduate students, principally in science. Also, I am trying to keep my explanations accessible to allow the general reader to understand modern science as well as place modern science in the context of the philosophy of science, whose ideas and theories about how science works span millennia. Philosophers of science, I hope, will be interested in my descriptions, so I will meet them in their own domain.

As a telling illustration of the importance of the topic of scientific truth and as a prediction of what might happen based on "the science," I offer the following example, where even highly trained people (in the humanities) struggle with what science can offer and the meaning of probability. At the UK COVID-19 Inquiry hearing on 8th November 2023, Sedwill, Keith, and Hallet discussed the following:

> Lord Sedwill (Head of the UK Civil Service in 2020): there was too much focus, including in the briefings to Cabinet, on the reasonable worst case rather than from the deep experts, "Here's what I think will happen."
>
> LADY HALLETT (Chair of the Inquiry): Yes, exactly.
>
> MR KEITH (King's Council i.e. the questioning lawyer): And it's the same dichotomy, isn't it, reflected in that briefing in the second COBR[1] from the Chief Medical Officer? He says: "This is the reasonable worst-case scenario, however in reality the real scenarios are the following..."

NOTE

1. The Cabinet Office Briefing Rooms (COBR) are meeting rooms in the UK government's Ministerial Cabinet Office in London. These rooms are used for committees which co-ordinate the actions of government bodies in response to national or regional crises, or during overseas events with major implications for the UK. The meetings are popularly referred to as COBRA meetings, including in the media.

REFERENCE

Sedwill, M., Keith, H. and Hallet, H.C. (2023) at the UK Covid 19 Inquiry November 8th 2023, p. 13. https://covid19.public-inquiry.uk/wp-content/uploads/2023/11/08193313/2023-11-08-Module-2-Day-20-Transcript.pdf

Author Biography

John R Helliwell, DSc (Physics, University of York), DPhil (Molecular Biophysics, Oxford University), is Emeritus Professor of Chemistry at The University of Manchester, where he served as Professor of Structural Chemistry from 1989 to 2012. Academic teaching from 1979 to 1988 was at the Universities of Keele and York in the physics departments there. He is a researcher in the fields of crystallography, biophysics, structural biology, structural chemistry, and data science. He was also based at the Synchrotron Radiation Source at the UK's Daresbury Laboratory in various periods of appointment between 1979 to 2008, including in 2002 as Director of Synchrotron Radiation Science. He is a Fellow of the Institute of Physics, the Royal Society of Chemistry, the Royal Society of Biology, and the American Crystallographic Association, and an Honorary Member of the British Crystallographic Association and of the British Biophysical Society. He is a Corresponding Member of the Royal Academy of Sciences and Arts of Barcelona, Spain, and an Honorary Member of the National Institute of Chemistry, Slovenia. His awards include the European Crystallographic Association Eighth Max Perutz Prize 2015, the American Crystallographic Association Patterson Award 2014, and the "Professor K Banerjee Endowment Lecture Silver Medal" of the Indian Association for the Cultivation of Science (IACS) 2001. He has published over 200 scientific research papers and several books, e.g., *Macromolecular Crystallography with Synchrotron Radiation* with Cambridge University Press (1992), published in paperback in 2005, and *Macromolecular Crystallization and Crystal Perfection* with N E Chayen and E H Snell), Oxford University Press, International Union of Crystallography Monographs on Crystallography (2010). He has published several *Scientific Life,* popular science, books in recent years, which are with CRC Press, Taylor & Francis.

He has served in roles of major responsibility such as President of the European Crystallographic Association (2007–2010), Chairman of the

International Union of Crystallography's (IUCr) Commission on Journals (1996–2005), and Chairman of the IUCr Diffraction Data Deposition Working Group (2011–2017) and its Committee on Data (2017–2023) as well as its Representative to the International Council for Scientific and Technical Information (ICSTI; 2005–2014) and the International Council of Science's Committee on Data "CODATA" (2012–2023). In the past thirty years, he has chaired several international advisory committees for synchrotron, and more recently, neutron, facilities' development and their users' science. He was Leader of the UK Delegation at the International Union of Pure and Applied Biophysics Congress and General Assembly in New Delhi, India, in 1999, and was Leader of the UK Delegation at the International Union of Crystallography Congress and General Assembly in Prague in 2021.

Scientists, in their career development, seek guidance. He has been a senior mentor at the University of Manchester, both in the Department of Chemistry and in its Manchester Gold scheme, as well as a director in the scientific civil service, as mentioned above.

He has had interests in the philosophy of science for a long time. So much so that he has written quite a number of book reviews on the topic in his (semi-)retirement since 2012. He has also reviewed books on how science is communicated and how it is managed for the public good, as well as books on scientific career development.

Other books by John R Helliwell in his Scientific Life Series

Skills for a Scientific Life
The Whys of a Scientific Life
The Whats of a Scientific Life
The Whens and Wheres of a Scientific Life

CHAPTER 1

What Are the Main Themes of the Philosophy of Science in Understanding Science's Efforts to Reach Truth?

As a practicing scientist specializing in the science of crystallography, which spans many fields of research, I have a picture in my mind of what we scientists are all trying to do but which is necessarily subjective. It is a picture framed by my upbringing, my place as a member of a particular country, the United Kingdom, and the research roles I have taken up. Philosophers of science view what we as scientists do from the outside, and perhaps therefore, they may have a greater level of objectivity to our aims and practices as scientists. So, what are the main themes of the philosophy of science in understanding science's efforts to reach truth? The philosophers of science seem to see themselves as being within a branch of philosophy rather than science. Let's take a few examples of how philosophers of science see themselves.

Firstly, I quote from a lecture by Professor Helen Longino of Stanford University (https://www.youtube.com/watch?v=631gObE7ctA)

DOI: 10.1201/9781003405399-1

> Philosophy of science is an interpretive and critical engagement with the sciences.

Secondly, Sir Freddie (A J) Ayer, in his autobiography *Part of My Life* (Ayer 1977), describing his school education at Eton College, UK, wrote that he had very little schooling in the sciences. As an example, he recalled only a brief encounter with Bunsen burners in a chemistry lab. Trying to be positive about this, I would say it is a bit like seeing a new Minister of Science appointed in the UK without any obvious knowledge of science. So, a prime minister in charge of ministerial appointments would stress that there is no bias one way or the other of such an appointee in considering funding allocation decisions for the UK sciences. Nevertheless, as a scientist, I find such an approach to be based on ignorance, and rather difficult to accept. That said, I found empathy in reading the publications of Sir Freddie Ayer. These included, besides his autobiography, his book entitled *"Probability and Evidence."* My reading of Ayer led me to what I imagined would be a particularly promising book: von Mises (1957), 3rd edition, with the promising title *"Probability, Statistics and Truth."* However, truth as a word does not feature in the subject index, or in the contents pages of von Mises (1957). Instead, at page 168 von Mises writes "*The problem of the existence or non-existence of a 'true' value of every quantity belongs to the realm of epistemology and we need not be concerned with it here.*" I would also mention that von Mises does not make the distinction between systematic and random errors. With one method alone that value may have precision, but it requires at least a second method, i.e., with its different systematic errors, to establish the accuracy or not of either method. A scientist practicing both those methods would of course try to eliminate or at least reduce the systematic errors in each.

Thirdly, in terms of books, a most helpful book on the whole topic of the philosophy of science I find is Cover, Curd, and Pincock's (2012) book, an edited multi-author volume entitled *"Philosophy of Science: The Central Issues."* This magnum opus comprises 1393 pages in its 2011 2nd edition. It is in this book that I found Helen Longino again and her chapter entitled "Values and Objectivity" which I again empathized with. I immediately liked her text:

> Often scientists speak of the objectivity of data. By this they seem to mean that the information upon which their theories and

> hypotheses rest has been obtained in such a way as to justify their reliance upon it.

Of the 126 terms defined in Cover, Curd, and Pincock's (2012) Glossary, the one that I think exactly captured the sciences is the one for scientific realism, which is:

> generally taken to be the doctrine that the world studied by science exists and has the properties it does independently of our beliefs, perceptions, and theorizing.

Having found a comfortable location for my own thinking within the pantheon of philosophers of science, as a realist, I then find out that there is a countermovement known as **anti-realism.** In this, nicely described by Samir Okasha (2016), we find that this is:

> that the aim of science is to find theories that are empirically adequate, i.e. which correctly predict the results of experiment and observation, not truth.

In my own laboratory's studies of protein ligand binding, our research has sought a prediction of the binding affinity of a ligand for a protein binding site from its 3D molecular structure (Bradbrook et al. 1998; for an update and overview, including its continued failure to be achieved so far, see Helliwell (2022)). The alternative to such predictability (and I am tempted to write simple predictability) is brute force surveying of many thousands of molecular fragments to build up a reasonable picture of what binds to a protein binding site and what doesn't. Thus, I can see that the empirical adequacy of brute force ligand screening is the goal of such an approach, although, it is less elegant than a theoretical prediction. Also, the brute force method, albeit currently a more certain approach, is costly versus an in-silico method of finding out increased affinity binders of a ligand at a protein binding site based on thermodynamics. I discuss these two approaches in more detail below.

In relating the types of scientific progress to actual, historically important scientific advances in crystallography in the past 100 years or so Massera and Helliwell (2023) highlighted ten "Golden Oldies" publications. In conclusion, we offered the following summary:

> The examples show, to a greater or lesser degree, that the context of community developments and thinking are important to the individual researchers that we have highlighted. In more recent times, we can refer to the community-agreed validation reports of checkCIF and of the PDB as examples of community consensus. In our highlights there are examples of science advance by incremental change, albeit fairly large increments. There is at least one paradigm shift, most notably the first X-ray crystal structures determined by W. L. Bragg (1913). In these matters then we can see that crystallography fits the different theories of the philosophy of science of how science advances: increments, paradigm shifts and consensus as well as team play and the insights of individuals.

The Kuhn paradigm shift is a famous theory in the philosophy of science about how science advances (Kuhn 1996). In crystallography, the transition from studying the external faces of crystals that predated the discovery of X-rays in 1895 to the X-ray crystal internal structure analyses from 1912 onwards is to my mind a clear paradigm shift. A second example James Kaduk (personal communication), in commenting on my book draft, remarked: "*In the 1960s plate tectonics was just starting to be accepted. Some of my early textbooks provided explanations which did not seem to make sense, and plate tectonics provided an elegant new explanation for many observations. Plate tectonics was a true paradigm shift, showing that the previous consensus was wrong.*"

Massera and Helliwell (2023) also thought that another aspect to be emphasized was that:

> a challenging aspect was to convey to you, the reader, the context of the time when the research was undertaken, and when the articles were published. We hope that we have succeeded to a degree and have used this phrase 'science pull and technology push', where obvious, as a simple starting clarification of the context of research at the time.

A key feature of "truth" in science is the reproducibility of what has been reported in a publication. To assess reproducibility, the underpinning data must be included with the narrative of an article. Following on from this, others can design their own experiments to try to replicate the first publication's results. This figures highly in a scientist's mindset. I explore it in detail in Chapter 5 of this book. Also, in Chapter 3, the measuring of the right thing is emphasized and within that ideal, the approach adopted by a scientist might well have to be pragmatic. This does mean that there can be layers of truth as successive approximations to the functioning reality of a system of scientific interest.

BIBLIOGRAPHY

Ayer, A. J. (1977) Part of My Life: The Memoirs of a Philosopher. 318 pages. Harcourt, Brace and Jovanovich, New York.

Bradbrook, G. M., Gleichmann, T., Harrop, S. J., Habash, J., Raftery, J., Kalb (Gilboa), A. J., Yariv J., Hillier, I. H.& Helliwell, J. R. (1998) X–ray and molecular dynamics studies of concanavalin A glucoside and mannoside complexes: Relating structure to thermodynamics of binding. Faraday Transactions 94(11), 1603–1611.

Cover, J. A., Curd, M. J. & Pincock, C. (2012) Editors of "Philosophy of Science the Central Issues". 2nd edition. 1393 pages. W. W. Norton & Company, New York.

Helliwell, J. R. (2022) Relating protein crystal structure to ligand-binding thermodynamics. Acta Cryst F 78, 403–407.

Kuhn, T. S. (1996) The Structure of Scientific Revolutions. 3rd edition. University of Chicago Press, Chicago.

Massera, C. & Helliwell, J. R. (2023) Golden oldies: Ten crystallography articles that we think must be read. Acta Cryst. E 79, 580–591.

Okasha, S. (2016, July 28) Philosophy of Science: A Very Short Introduction. 2nd edition. Very Short Introductions (online edn, Oxford Academic, Oxford). https://doi.org/10.1093/actrade/9780198745587.001.0001, accessed 30 March 2023.

von Mises, R. (1957) Probability, Statistics and Truth. 3rd edition. Dover Publications, New York. Copyright Geroge Alln and Unwin Ltd.

REFERENCES

Longino, H. Lecture entitled. Philosophy of science is an interpretive and critical engagement with the sciences. https://www.youtube.com/watch?v=631gObE7ctA

Longino, H. (2012) Chapter entitled "Values and Objectivity". In Cover, J.A., Curd, M. J. and Pincock, C. (Eds.), Editors of "Philosophy of Science the Central Issues". 2nd edition. 1393 pages. W. W. Norton & Company, New York.

CHAPTER 2

An Atlas of Scientists' Subject Areas in Seeking Scientific Truth

INTRODUCTION

In this part of the book, I explore the different disciplines of science, namely the approaches and methodologies of scientists for realising scientific truth in physics, chemistry, the pharmaceutical sciences (closely allied to chemistry), biology, computer science, and mathematics. Firstly, I organize this part into common themes and then, secondly, describe the specific approaches adopted by the scientists in these specialized domains as well as offer comments on the social sciences. There is a good connection between the common base and the philosophers of science analyses of what is being done by scientists, which I described in Chapter 1. Physics as a subject was originally called "natural philosophy," and it is then the subject with the most commonality with philosophy of science. Then, thirdly, I describe standards of measurement.

THE COMMON APPROACHES TO SCIENTIFIC TRUTH ACROSS THE DISCIPLINES

All the subject areas of science have experiments and theory specialists. Usually, the theoretical scientists are explicitly labelled as such. So, there will be, in an academic department, a professor of theoretical physics or

DOI: 10.1201/9781003405399-2

theoretical chemistry. But then the professor of physics or chemistry is not usually labelled as a professor of experimental physics or chemistry but might be categorized by a title such as professor of condensed matter physics or inorganic chemistry. So, this classification of the theorists is clearly important. The distinguishing features of theory-driven scientists might be immediately clearer if one considers, say, Albert Einstein. He was a theoretical physicist who, by simply posing a question, such as "what if the speed of light is finite?" then works out with pencil and paper what the consequences would be. It is not known if those predicted consequences are true until verified by experiment. There is a commonality in the role of theorists in any given science subject, the simplest way being the asking of a question or forming a hypothesis. Experimentalists proceed in the same way with a hypothesis and gather data to explore it. Confirming the predicted consequences of a theory made by a theorist will require appropriate experiments and appropriate equipment for any given science subject. But the experiments, even within that diversity, will have the commonality of establishing a basis of evidence, such as measured data from an apparatus or the yield of a compound from a chemical reaction. The assembling of empirical evidence and, as objectively as possible, assessing the match of that evidence to a theory's predictions are a form of the scientific method to advance knowledge. Often, one sees the words "the scientific method" as if there is one way of doing science to discover things. With complex systems or vast data sets, it can be very difficult to assess and make predictions. However, instead, to look for patterns in the vast data sets can be a good approach. This is an approach referred to as artificial intelligence and/or machine learning. A recent example of this is finding the solution to the prediction of protein folds from patterns within the databases of gene sequences and the three-dimensional protein structures in the Protein Data Bank, which is discussed later. A well-known example of discerning the details of the behaviour of the chemical elements and their chemical reactivities, as well as their physical properties, is the periodic table of the chemical elements discovered by Mendeleev in the 19th century. Obviously, this was achieved in the pre-computer, pre-digital age. The first digital computer was the Colossus II at Bletchley Park, invented by Alan Turing and created by electrical engineers with the purpose of code breaking in the Second World War.

All areas of science, in their proper elaborations at least, seek reliability of their results so that those results can be built upon. An individual study must provide all the underpinning data so that any subjective choices made by a researcher can be scrutinized. It is also becoming possible to preserve all the data and not just the portion selected by a team for publication. So, the workflow of those choices can be made clear along with the data at each step. The three core stages between steps start with the raw data, also known as the primary data. These are processed, also known as data reduction, and there are likely to be multiple software packages available and these will likely have different stages of development called versions. A careful track must be kept of the versions, and those software versions must be archived. A model fitted to the processed data is called the derived data stage. The best fit to the observed, processed data is monitored with a calculation from that model, and the difference between them is the residual. The smallest possible residual is the conclusion of those calculations and that best fit forms a measure of the degree of precision. To achieve accuracy is more difficult. Each method has its own systematic errors, which the experimentalist attempts to eliminate as much as possible. A second method though is needed to measure the same entity, and it will have its own systematic errors. The combination of two or more methods, each with its own systematic errors, leads to a best fit in each case. The models for each method can be compared. Examples of this from my own field of structural biology will be described in Chapter 4. A noteworthy case from my own research is the understanding of the molecular basis of coloration in marine crustacea, where X-ray crystallography, ultraviolet visible light spectroscopy, and solution X-ray scattering formed an integrated approach.

All areas of science have uncertainty in their results. This is well understood by scientists, and statistical and probabilistic methods are routinely applied by them. These mathematical assessments can be quite complex in their attempt to rationalize the measured data. Quantities that are offered without an uncertainty estimate are obviously problematic. The scientists may appreciate the challenges in providing these, but the consumers of scientific results may have little to no notion of the certainty of a quantity but can be guided by estimates of uncertainty when guided by the scientists involved. Politicians seem to take results as certainty. It also seems that politicians can, however, attribute their own uncertainties to results that disagree with their beliefs; climate change scepticism or vaccination reluctance are examples. The political landscape and cultural attitudes can also vary from region to region and country to country around the world. These differences can influence how politicians approach and prioritize scientific truth in their decision-making in different regions. Chapter 9 of this book will describe instances of these and where the terms misinformation or even deliberate disinformation can be considered.

Specific subject area approaches to scientific truth

(i) Physics

Physics has a firm view of the errors that occur in any given experimental measurement using a method, and introduced this terminology, described above, where the errors are of two types: random and systematic. Best efforts are made to minimize the systematic errors in a method. An estimate of the standard uncertainty is made of a parameter value that was the object of the measuring procedure. The aim is to achieve as precise a value or model as possible. In this sense, the true value or true model is never reached. The use of a second method, which will have its own systematic errors of measurement but different ones from the first method, will also aim to achieve as precise a value of a parameter or model. The two methods used in this way together mean a cross-check can be made of the outcomes, and thereby an assessment can be made of the accuracy of the parameter or model. These exact definitions used by physicists are very often not adhered to in other areas of science, which can use the words precision and accuracy interchangeably. This loosening of the definitions can even go so far as referring to a prediction as accurate even though the best that can be done is that a prediction can only give a confidence estimate of its prediction.

(ii) Chemistry

Chemistry compared with physics must immediately wrestle with complexities. Whereas physicists can create idealized situations that are often treatable with exact analysis, this is not the case with chemical systems. In the periodic table of the elements, only the hydrogen atom can be treated in a physics and mathematically exact way to calculate its spectral properties. This is done using the Schrodinger wave equation based on the interesting but curious behaviours of an electron as having wave properties as well as in other circumstances behaving as a particle. For elements beyond the hydrogen atom in the periodic table, the calculation of an element's spectral properties requires approximate calculations referred to as perturbation theories. Beyond the individual elements the focus of chemistry is on making a vast number of possible chemical compounds from them.

A practical means of establishing the truth of a researcher's understanding of a chemical system can be monitored by the product yield of a chemical reaction.

In the training of chemists in their degrees, as well as an experimental and theoretical sub-division, there are further sub-divisions into inorganic, organic, and physical chemistry. Alongside these, the practical classes are split into chemical synthesis on the one hand and measurements on the other. The role of apparatus and data for all sorts of physics-based methods is vital for new chemical compounds' characterisation.

(iii) The pharmaceutical sciences

This scientific field is closely interfaced to chemistry. However, new drug compounds from organic or inorganic synthesis, characterized by the methods of physical chemistry (strongly rooted in physics), must then be assessed for medical or veterinary treatments by first doing clinical trials. There are stringent ethical considerations. So, strict procedures are enforced by regulatory agencies. In the UK, for example, the National Institute of Clinical Excellence was established in 1999 to provide evidence-based guidance and recommendations for healthcare interventions, treatments, and technologies within the National Health Service in England and Wales.

(iv) Biology

In Chapter 4, I go into considerable depth about structural molecular biology, increasingly referred to as structural biology as its research domain expanded from the molecular level to the cellular level. In Chapter 4, I also discuss the interface with holistic biology. Biology is a subject of vast complexity and has many different domains. The book by Ernst Mayr, "*Why Is Biology Unique?: The Autonomy of a Scientific Discipline,*" gives a distinctive view of the domains of biology and how they interface one with the other and externally with other subjects. New discoveries often arise from careful observations of living organisms and their behaviours in their natural environments. Perhaps most famously, the voyage of Charles Darwin from 1831 to 1836 to Tierra del Fuego in South America allowed him to make extensive 'field studies' and which led him to the theory of natural selection.

(v) Computer science

This has been a fast-growing area of science and is popular with an ever-increasing number of undergraduate students. The topics that fall within its domain include programming in "up-to-date" languages. Very topical, though, is artificial intelligence and machine learning, whereby patterns in data that may be missed by a human can be surveyed by computer. These are also leading to improved query tools, compared with Google web searches, like ChatGPT (https://chat.openai.com/auth/login). However, as with internet searches made by a human, the results from a ChatGPT query must be scrutinized for their sense by that human. The difficulty is that that human may not be well informed or capable of exercising critical judgement on a ChatGPT answer. Therefore, there is still very much a need for variance estimates on the truthfulness of the outputs of such artificial intelligence and machine learning tools.

Within open science as a headline theme, open software source code is a particular and very important category. Given that the "first stored program computer" is as recent as 1948 ('The Manchester Baby' https://en.wikipedia.org/wiki/Manchester_Baby) the incredible growth of computing power in the decades since has changed all our lives. A measure of this growth would be Moore's Law of a doubling of computer memory every two years. So, it is now routine for a computer laptop to have a multigigabit central processing unit and likewise for its solid-state memory.

(vi) Interdisciplinary subjects

These can be independently organized or deemed sub-domains of a given single science subject. Examples of the former would be biochemistry and of the latter would be biophysics. But there is no hard rule. A subject like crystallography, my own, does not have its own undergraduate academic department, but it does have one example of its own postgraduate department, based at Birkbeck College, University of London. Crystallography has professional organisations representing crystallographers at national, continental, and international, i.e., global levels. Likewise, biophysics is similarly organized. There is no particular hallmark to their views of scientific truth than single science subjects.

(vii) Mathematics

Within the sciences mathematics is vital. Closely allied to pure or applied mathematics are the domains of data science, statistics, and probability as mentioned in various contexts, across the sciences, above.

(viii) Standards of measurement

The training of students in schools and universities promotes standards of measurement in practical classes as a complement to their classroom training. Ultimately, the absolute reference point involves government agencies such as the National Institute of Standards and Technology in the USA and the National Physical Laboratory in the UK. These agencies promote and maintain standards of measurement in science and technology. The science of measurement is known as metrology. A vital component is the understanding of the measurement errors that always arise, both systematic and random, as I have described above, and so just how they should be treated and presented in proper and uniform ways is vital. The variability of the standard of understanding and reporting errors is a widespread weakness across the science disciplines. There is also a widespread mispractice where false precision (too many decimal places) of measured values might be offered by a scientist or even no estimate given at all of the uncertainty of a measured value. This is an aspect where the truth of scientific results can be badly interpreted by politicians and the public. It also gives opportunities for those who wish to misinform the public, or even worse, disinform the public. I deal with these aspects in Chapter 9.

(ix) Social sciences

The International Science Council was formed a few years ago as a merger of the International Council for Science and the International Social Sciences Council. My expertise is in the physical, chemical, and biological sciences and not the social sciences. However, I welcomed this merger. The International Union of Crystallography (IUCr) was lukewarm to the proposed merger. I thought this was a mistake, and so in 2017, I stood for election as President of the IUCr. In the 4th and final portion of my speech to the 2017 IUCr General Assembly held in Hyderabad, India, I spoke as follows:

> Currently the International Council for Science and the International Council for Social Sciences are proposed to merge. We crystallographers have broad science perspectives as well as a long history of societal contributions based on our science. A heroine of mine is Dorothy Hodgkin who contributed greatly to such matters. My DPhil supervisor was Margaret Adams, and she was Dorothy's supervisee. It was a privilege to interact with Dorothy in that environment. I believe then that the IUCr should join in enthusiastically with this ICSU and ICSS merger and promote it. In the ACA Transactions Symposium held in Philadelphia in 2015 on Crystallography and Sustainability[$] I set out my vision in my talk for how we in crystallography can help develop the United Nation's Millennium Sustainability Development Goals. I would also wish to continue to promote this effort as IUCr President.
>
> *$Helliwell, J.R. (2015) pages 8 to 19 of:-*
> *https://www.amercrystalassn.org/assets/volume45.pdf*

As it turned out in the voting, I was third in a poll involving five candidates.

Social scientists, I believe, follow a quantitative methodology. I was very impressed at the 2018 International Data Week in Botswana where I learnt about their FACT adherence to their data (van der Aalst et al 2017). They saw this as needed as well as FAIR data. I describe this in detail in Chapter 5 of this book. In observing social scientists, I note that they navigate the complexities of human behaviour while striving for objectivity, ethical considerations, and engagement with diverse perspectives. I also know that in my own research work, especially at the Synchrotron Radiation Source, our efforts with the Daresbury Analytical Research and Technical Services (DARTS) offered societally relevant research from the physical sciences. The interface between these different areas of science is vital for the future. Could this be controversial? In my role as Director of Synchrotron Radiation Science at the Council for the Central Laboratories of the Research Councils (CCLRC), I attended Senior Management Development Programme Training (see Figure 2.1). A director of another UK national facility spoke strongly against societally relevant research, calling it "work that was like taking in washing." This was as if it was only what impoverished people do. Naturally, as an enthusiast for this type of research, where I attended the monthly DARTS meetings (see Figure 2.2) and had done the costings for the pharmaceutical industries using protein crystallography Synchrotron Radiation Source (SRS) beamlines, I challenged this view. This provoked a furious row between us. So, not only in the IUCr but also in the CCLRC I had encountered a prejudice against societally relevant research.

FIGURE 2.1 A copy of the front cover of my November 2002 Senior Management Development Programme training manual.

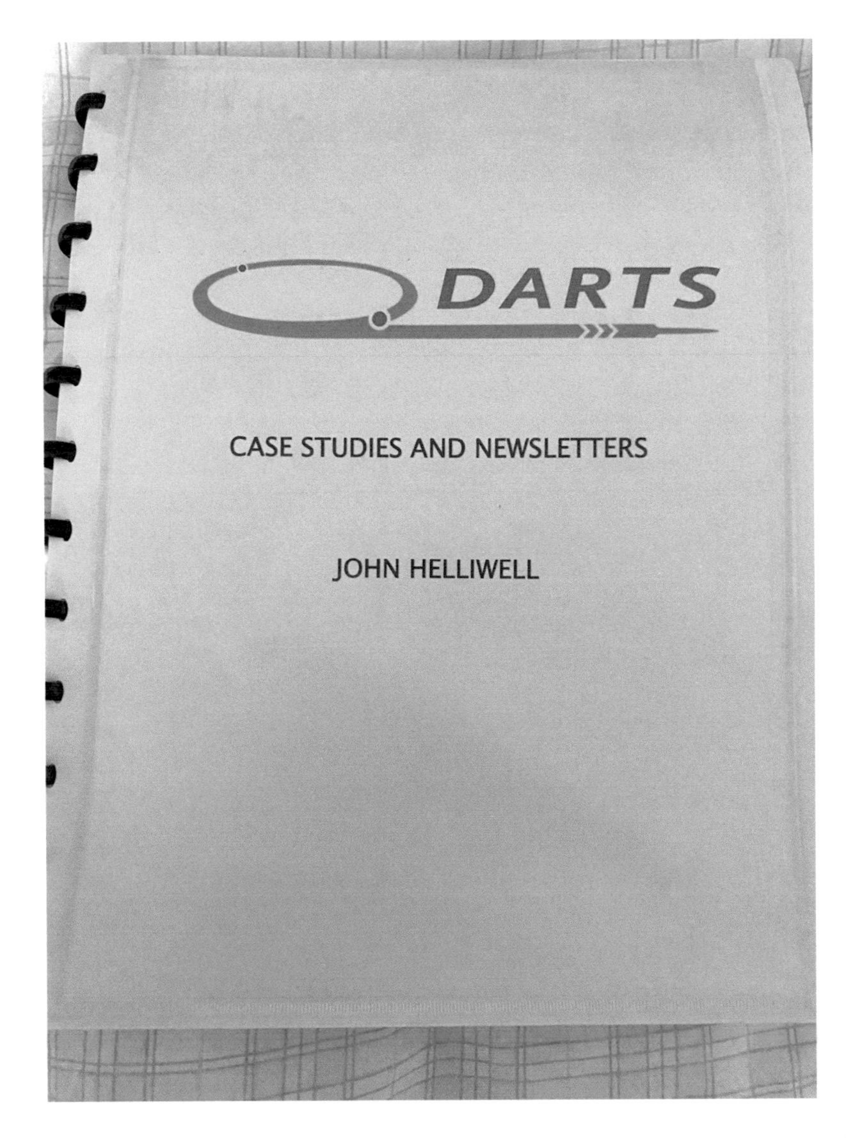

FIGURE 2.2 My Director's copy of the Daresbury Analytical Research and Technical Services Case Studies, which I found useful for showing visitors over a coffee or a tea.

To conclude this Chapter 2, across the whole atlas of science subjects, I quote Isaac Newton (1643–1727):

> I have been like a boy playing on the seashore diverting myself in now and then, finding a smoother pebble or a prettier shell

> than ordinary, whilst the great ocean of truth lay all undiscovered before me.

I think that all scientists across the domains that I have summarized in this atlas see a great ocean of undiscovered truth in front of them. Their curiosity is a big driver to explore the truths in those great unknowns.

CHAPTER 3

Measure the Right Thing. What Is the Best Probe of the Structure of Matter?

What Is the True Value? Precision and Accuracy

THERE ARE VARIOUS PRACTICAL details impinging on the ideal of scientific truth. These are treated in the numerical sense in the field of probability and statistics. But we should also consider layers of truth and being forced into measuring far from the ideal due to the need for pragmatism. This topic is at the interface of philosophers of science descriptions and scientists' efforts in the laboratory. As one example of this approach, in structural biology, integrative techniques are used to span all the relevant length scales of living systems, from atoms to molecules, from cells to organs, and to the whole organism (Ward et al. 2013). I address this as a case study in Chapter 4.

In **measurement physics,** we must know what the difference between precision and accuracy in science is. I described this from the point of view of physics in Chapter 2. In my own research area of crystal structure

DOI: 10.1201/9781003405399-3

analysis, i.e., molecular-level structure determination, of course, we need probes. For diffraction, these are X-rays, electrons, and neutrons. In spectroscopy, we have a major player as a probe in the method of nuclear magnetic resonance (NMR). In terms of the sample to be studied, these probes can be used to investigate all the states of matter, such as a crystal, a fibre, an amorphous solid, a liquid, or a gas. The probes themselves can cause damage to the sample depending on the dose used. The required dose to obtain the information of interest goes up the smaller the sample volume. The least damaging probes in diffraction are neutrons. Electrons are an interesting case whereby, through the use of low doses in microscopy and by studying many copies of single particles of proteins or complexes of proteins on an electron microscope grid, marvellous detail can be obtained without needing a crystal sample. If a sample gets too thick, however, the strong electron interaction with matter can become problematic. An interesting middle ground between single particles and a "regular sized" crystal (~100 microns) is the micron, or less, single crystal size, whereby the use of electron diffraction can still be very effective; this domain is known in my research field as "microED," i.e., micro electron diffraction. I will explore these probes of the structure of matter in more detail shortly.

W Lawrence Bragg (1968) in Scientific American wrote on X-ray crystallography that:

> "I have often been asked." *Why are you always showing and talking about models? Other kinds of scientists do not do this." Bragg answered that* "The investigator seeks a structural plan, a map that shows all the atoms in their relative positions in space; a list of spatial coordinates is all that is needed to tell the world what has been discovered."

What we see in an X-ray diffraction experiment is electron density, and we interpret it as atoms using our prior/additional knowledge. Like using X-rays, with neutrons we see the nuclei of atoms and with electrons we see the electrostatic potential map of a molecule. So, with each probe, we can see atoms with crystallography, *i.e.*, Bragg's map of all the atoms. We now must also ask: when are the atoms that we place in our electron density maps from X-ray diffraction, nuclear density maps from neutron diffraction, or electrostatic potential maps from electron diffraction or microscopy in the correct positions to explain a function? We need either an assay of function or a complementary technique providing a consensus or a predictive force from our crystal structure (e.g. site-directed mutagenesis of amino acids in a protein where we can change the function, such as the rate of an enzyme reaction based on its three-dimensional structure). These considerations drive to the core words in science: **precision** and **accuracy** (Figure 3.1).

When one method's measurements and a model's interpretation correctly match, this equals precision. When a second, different, method is used, its measurements lead to a second model, which can also be precise in its own way. But when both, or more, methods agree, we reach accuracy. There are other common words in measurement science to be mentioned: systematic error and random error. With any of the methods, both types of errors occur. One strives to remove systematic errors completely, if possible. Also, in terms of basics, the core point is that, as the International Union of Crystallography (IUCr) Working Party on Expression of Uncertainty in Measurement reported (Schwarzenbach et al. 1995), *"a measurement result is considered complete only when accompanied by a quantitative statement of its uncertainty."*

These terms have been considered from the electron charge density point of view by Sanjuan-Szklarz et al. (2020). The central issue is the use,

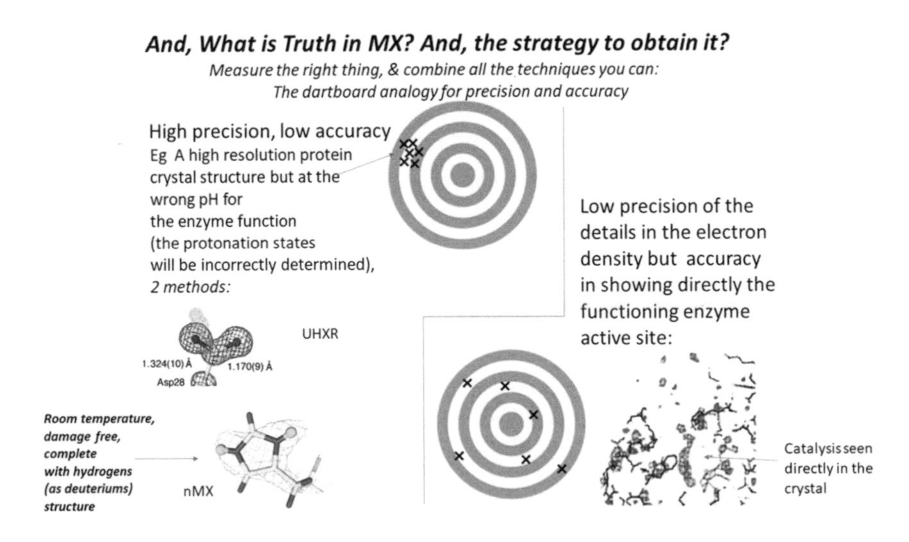

FIGURE 3.1 Precision and accuracy illustrated using a dartboard and set in the context of crystal structure analysis of proteins (macromolecular crystallography, 'MX'). Based on Helliwell (2022) lecture at CCP4 2022 for which a recording is available here: https://www.youtube.com/watch?v=I7e2EYrmBMg&list=PLrmG39_bWIGgvTugFLZaxHkR0WCkf8hQQ&index=24. UHXR is ultra-high resolution macromolecular crystallography and nMX is neutron macromolecular crystallography.

in general, of spherical electron densities to arrive at an atomic model rather than the physically more reasonable aspherical electron densities. As their Figure 6 nicely shows (Figure 3.2 here), the truth is not known, and so accuracy could, even should, be replaced by a reference structure, and the neutron crystal structure is their choice of reference.

There is a commonality between the approach of Sanjuan-Szklarz et al. (2020) and the one I describe here for protein crystallography, but since protein crystallography is at a lower, i.e., less good diffraction resolution than the one used by Sanjuan-Szklarz et al. (2020), the reasons to make the neutron protein crystal structure the reference instead are: the structural completeness with experimentally determined hydrogens for the ionisable amino acids, the use of physiological relevant temperature, and a protein model free of X-ray or electron radiation damage. That said, as Chemistry Nobel Prize winner Joachim Frank (https://www.nobelprize.org/prizes/chemistry/2017/frank/facts/) emphasizes the crystal itself is not the 'native state'. The crystallographer might well respond that the interactions of a protein with its neighbours in the crystalline array are nevertheless interesting and could well indicate the interactions experienced by the protein in a crowded living cell.

It is important to retain a big picture. So, in biology, we need the hierarchies of organisation and thereby the length scales of molecules, macromolecules, viruses, and organelles onto the complete organism, e.g., an animal, and even further, to whole populations and ecosystems. In a similar vein, we seek to understand the dynamics of molecular structures from femtoseconds to seconds and longer. These structural dynamics can include conformational change, atomic vibration, or bond making and

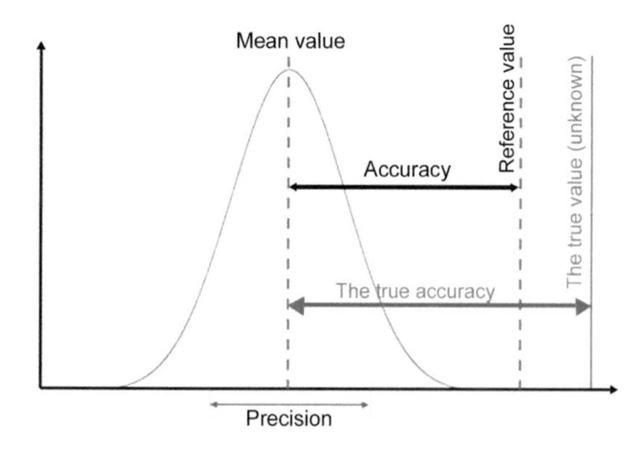

FIGURE 3.2 Definition of precision and accuracy (from Sanjuan-Szklarz et al (2020) with the authors' permission and of the IUCr Journals).

breaking. To this end, we purify molecules or complexes that in size, reach a limit at the several hundred Å scale. A different method known as electron tomography can view molecules in situ without extraction, such as coronavirus particles in a cell as described by Dr. June Almeida in her pioneering coronavirus studies (Almeida and Tyrrell 1967).

Today, crystallography's landscape of determining structures now coexists with DeepMind's prediction of a protein three-dimensional structure from a gene sequence, as seen in CASP14, Critical Assessment of Structure Prediction of Proteins (for a short resume, see Helliwell 2020). The Protein Data Bank Europe (PDBe) with DeepMind has established the AlphaFold Database (https://www.alphafold.ebi.ac.uk/), which provides open access to over 200 million protein structure predictions to accelerate scientific research.

Flexible proteins and complexes are resistant to crystallisation, and cryo-electron microscopy (cryoEM) has undergone great strides in resolution capability in recent years (see Kühlbrandt (2014) for an overview). For cryoEM, bearing in mind Joachim Frank's point above, freezing a single molecular complex is quite possibly more favourable than freezing a 1 micron protein crystal, thereby avoiding the possibility of cryoartefacts. In its results, cryoEM shows quite a range of resolutions across a structure, but so does crystallography with the atomic displacement "B" factors, some very large to the point that such a part of a structure shows no significant electron density.

These considerations bring me to a more detailed scrutiny of our probes of the structure of matter. Boris Vainshtein's (1964) book provides an excellent survey of our diffraction probes: X-rays, electrons, and neutrons. Neutrons give the nuclear positions, X-rays give the atomic electron density, and electrons give the atomic electrostatic potential.

PAIRWISE COMPARISONS OF OUR PROBES: ELEMENT IDENTIFICATION

The atomic scattering factor of atoms for X-rays steadily increases with the increasing atomic number. The resonance effects provided via the X-ray wavelength tuning to and around an individual element K or L absorption edge give X-ray diffraction an exquisite sensitivity to identifying a metal atom, especially bimetallic cases of neighbouring atomic number. The experiment with the most X-ray wavelengths, 11, to investigate a zinc-substituted gallium phosphate involving partial occupancy is that of M Helliwell et al. (2010). By tuning to each of the zinc and gallium K absorption edges, the precise locations and occupancies of these atoms were determined.

Nuclear scattering for different isotopes using neutrons also provides exquisite sensitivity with scattering contrast variation, such as hydrogen and deuterium. Likewise, there are element-to-element variations, but they are much less marked than using X-rays. A particularly attractive feature of the neutron-scattering factors is that deuterium is as good a scatterer for neutrons as carbon in magnitude. This is exploited in neutron protein crystallography to determine the protonation states of ionisable amino acid residues (aspartic acid, glutamic acid, histidine, and lysine).

Electrons, like X-rays, also show a steady increase in the electrostatic potential with increase in atomic number. Electrons are also much more sensitive to hydrogens than X-rays. For metals, a biological molecule known from chemical analysis to contain an electron-dense metal centre, the location of those metal atoms by electrons as probe can also be found. For bimetallic cases of closely similar atomic numbers, to identify them in position and type, is untested to my knowledge with electrons.

Both electrons and X-rays will very likely change the oxidation state of a metal. With X-rays; however, there is the important exception of a femtosecond-time-range pulsed X-ray laser, where the "diffract before destroy" approach (Neutze et al. 2000) is applicable.

SAMPLE SCATTERING POWER

Perhaps surprisingly, a definition of crystal sample scattering power was not introduced as an equation until Andrews et al. (1988); see their Table 3 comparing microcrystal diffraction of both chemical and protein microcrystallography. A different approach was adopted by Henderson (1995), namely that for naturally occurring biological material, electrons at present provide the most information for a given amount of radiation damage. The study by Henderson (1995) involved comparing X-rays, neutrons, and electrons. The results concurred with the Vainshtein (1964) approach of comparing the three probes and where electron scattering is the strongest. So, as Vainshtein (1964) observed, the very smallest crystals can still be studied with electrons. The obvious utility of the Henderson (1995) analysis is that for single particles of biological macromolecules, it encouraged the development of cryoEM. Once a crystal is grown to a size of ~0.1–1 micron, then the issue changes, namely, the electrons are overtaken by X-rays in their utility for structure analysis. Furthermore, if the crystal grows to ~1mm^3 or more then neutrons become viable and are by far

TABLE 3.1 Fundamental Comparison of X-rays and Electrons as Probes of the Structure of Matter. Source: Helliwell (2021)

X-rays	Electrons
• Yields electron density • Diffraction efficiency relatively low with respect to electrons • Mature method, well understood and validated (but still needs article with data files and checkcif report for submission of chemical crystallography articles to IUCr Journals[1]) • Phase problem • Radiation damage effects.	• Yields electron potential map • Diffraction efficiency high, good for very small samples, otherwise strongly absorbed and thereby: • Prone to multiple scattering, which can seriously affect bond distances and angles • CryoEM and electron crystallography still improving their capabilities as methods • No phase problem[2] • Radiation damage effects.

1. That procedure was extended to IUCr's biological journals in 2021.
2. In protein electron crystallography the phase problem is nicely discussed in Gemmi et al. 2019 and, as they summarize, molecular replacement using an already known protein structure, is so far the method applied.

the best probe because they yield a complete structure with hydrogens, at physiological temperature and are radiation-damage-free. Table 3.1 compares the fundamentals of X-rays and electrons, listing their pros and cons. Since neutrons are the ideal probe for crystallography, their use is purely one of the practical challenges, and these are compared with X-rays and electrons in Table 3.2.

There is one fundamental aspect of neutron crystallography: to analyze a neutron protein crystal structure first requires an X-ray crystal structure.

THE IMPORTANT ROLE OF NUCLEAR MAGNETIC RESONANCE

Nuclear magnetic resonance can determine protein structures in solution without crystallisation, and so, e.g., for an intrinsically disordered protein that won't crystallize, this is the only way forward. But the role of NMR has additional advantages when combining it with a crystal structure. Firstly, the NMR structure is an ensemble of structures fit to the NMR measurements. This means that where the protein polypeptide chain is particularly flexible and the X-ray crystal structure electron density has disappeared, the NMR ensemble fit still shows the range of positions for that portion. Secondly, in the relatively static core of a protein, as seen by the X-ray crystallography, the NMR data, still shows the dynamics.

TABLE 3.2 Practical Challenges of X-rays, Electrons, and Neutrons as Probes of the Structure of Matter.

X-rays	Electrons	Neutrons
• Radiation damage such as splitting of disulphides, truncation of amino acid side chains, changes to oxidation states of metal atoms/ions. These effects lead to use of cryo-temperature to (partially) mitigate them.	• Very strong interaction with matter, which is an advantage for very thin/small samples such as a single molecule or a nanocrystal • Bigger crystals have strong electron beam absorption/multiple scattering, and any bigger sample cannot be used with an electron beam, at least in transmission. • Careful scrutiny of the error estimates on bond lengths and angles is needed.	• Weak flux, so: • Use as broad a bandpass of the emitted neutrons as possible, i.e., Laue diffraction • Use as long a mean wavelength as possible to increase scattering efficiency. • Reduce any background scattering so as to maximize signal to noise (i.e., in biology change hydrogens for deuteriums with their 80× less incoherent scattering) • Grow as big a crystal as possible, e.g., ~1 mm^3 (typical range 0.1 to 8 mm^3) • Maximize the full exploitation of perdeuteration of a protein.

There is a profound set of ideas lurking here amongst the different molecular-level methodologies. As crystallographers, we can be easily misled by our static pictures and forget that everything is moving. We see the long-range time and spatially averaged electron density and assume that all unit cells in our crystal are the same. Nature/reality is more complex than that, so crystallography really is about generating **models** that explain the diffraction data. This was highlighted by Wuthrich and Wagner (1975) with respect to aromatic side chains such as phenylalanine, where distinctly different NMR spectra could be seen showing three possibilities: continuous spinning, 180-degree flip, or static. Thirdly the application of NMR to a crystal allows for both the ordered and disordered structures in the sample to be studied simultaneously. The book NMR Crystallography (Harris, Wasylishen, and Duer 2009) surveys many examples of this. Chapter 27 by Middleton (2009) in particular is devoted to structural biology applications. Middleton's Figure 27.2 (from Martin and Zilm 2003, Figure 3.3 here) shows the effects of sample preparation in solid state NMR comparing polycrystalline and nanocrystalline, which yield "practically identical" measurements, and thirdly, lyophilized ubiquitin protein shows poor resolution, indicating structural heterogeneity.

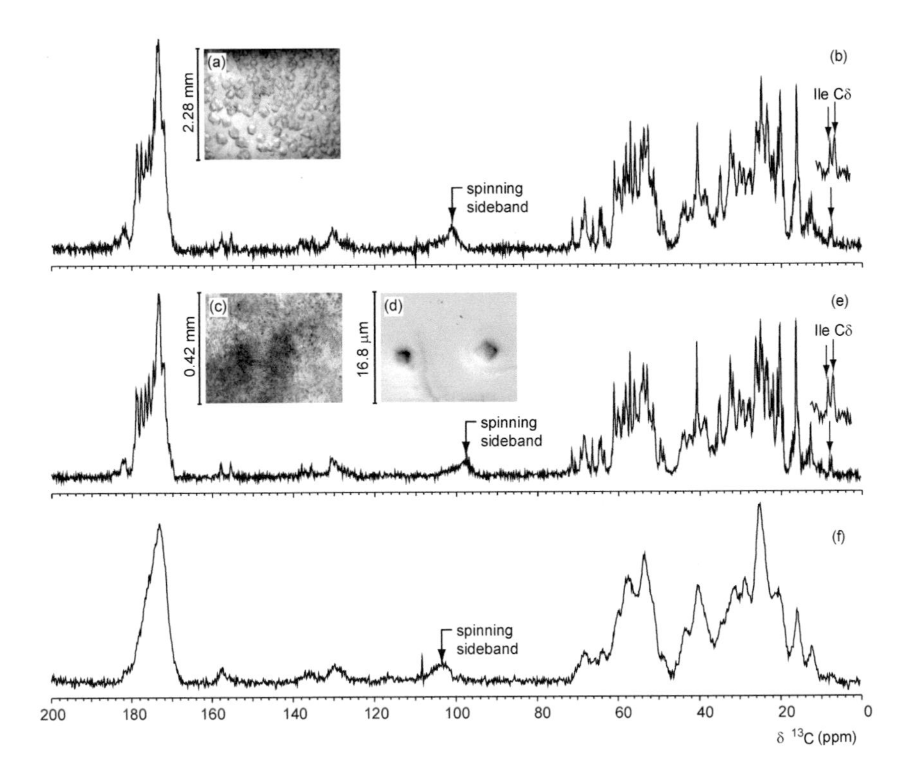

FIGURE 3.3 An example of the effect of sample preparation on the resolution of ^{13}C CP-MAS solid-state NMR spectra of proteins in the solid state. (a) A spectrum of polycrystalline ubiquitin composed of crystals grown in a sitting drop. Inset shows crystals at 5x magnification using a light microscope. The crystals measure approximately 200 μm across the widest point. (b) A spectrum of ubiquitin nanocrystals produced using rapid batch crystallization looks practically identical to that of the polycrystalline sample. Inset shows crystals at 255x magnification under a light microscope. (c) A spectrum of lyophilized (freeze-dried) ubiquitin shows poor resolution, indicating structural heterogeneity in the sample. (Reproduced from Ref. 11 copyright Elsevier, 2003 R.W. Martin and K.W. Zilm J Magn. Reason., 2003, 165, 162-174.).

Middleton (2009) also points out that where a crystal is available but does not diffract or shows limited diffraction, then NMR crystallography could be applied to study protein ligand binding on soaking with a ligand.

Lewandowski et al. (2015) used solid-state NMR to investigate the protein and solvent motions of nano- and micro-crystalline, fully hydrated protein "GB1" at temperatures from 105 to 280 K. GB1 is "a small globular protein specifically binding to antibodies." Their study describes a hierarchical change in dynamic behaviour (see especially their Figure 4; Figure 3.4 here) showing "*a unified description of the essential conformational energy surface, relating the amplitude and activation of solvent, sidechain, and backbone motions in a hierarchical distribution, as well as unambiguous identification of NMR line broadening at cryogenic temperatures.*" The key point here is that the truth of a structure and its dynamics depend on the measuring temperature of the probe, nicely illustrated in Figure 3.4.

Stepping outside protein examples Bryce (2017) provides a topical review of the use of NMR crystallography across a wide range of materials, spanning inorganics and organics as well as proteins. These various examples show the complementarity of NMR and crystallography in giving a more complete structure and dynamics understanding of all types of molecules. The IUCr established a Commission on NMR Crystallography and Related Methods in 2014 in recognition of these benefits (https://www.iucr.org/iucr/commissions/nmr-crystallography).

DIVERSE FURTHER FRONTIERS IN PURSUIT OF STRUCTURAL ACCURACY

There are diverse further important frontiers in the analytical three-dimensional characterisation of molecular structures. These include small-angle X-ray scattering, small-angle neutron scattering, mass spectrometry, X-ray absorption spectroscopy, infrared spectroscopy, and X-ray imaging.

I recall that, during my doctorate more than 45 years ago, the amino acid sequence of one's protein was not necessarily known, and special collaborations were established to determine the sequence. This is generally now all done via gene sequencing not protein–amino acid sequencing. Posttranslational modifications can occur, and in one's electron density maps, the missing electron density of amino acid side chains can raise anxiety about how to interpret those portions; debates such as on the protein crystallography bulletin board "CCP4bb" show the community view is split between letting the atomic displacement parameters inflate

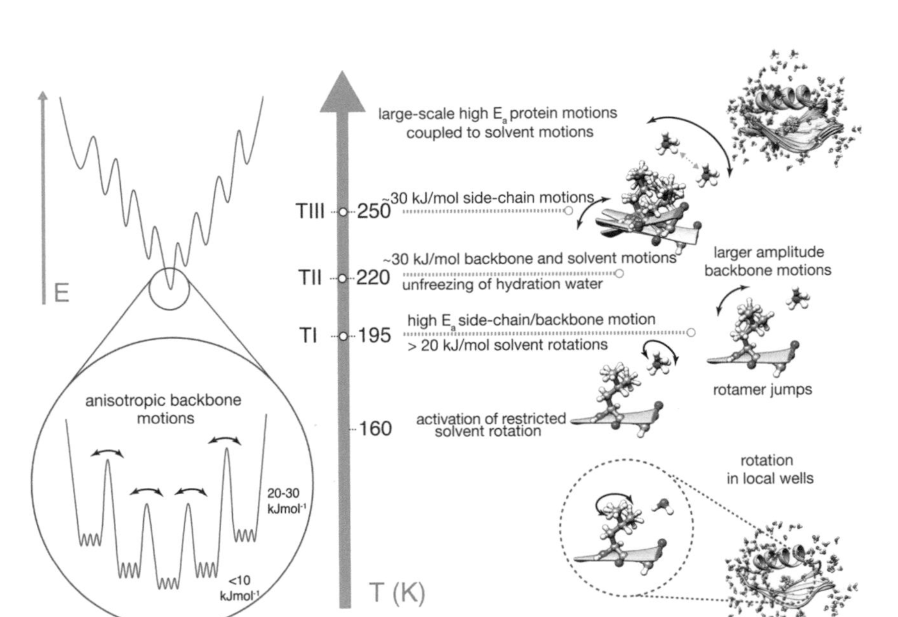

FIGURE 3.4 Summary of hierarchical dynamic behaviour of the protein-solvent system as observed by solid-state NMR in a microcrystalline globular protein GB1. The approximate temperature for the transitions between dominant dynamic modes is indicated on the vertical axis arrow. The image in the top right corner represents an ensemble extracted from a 200-ns molecular dynamics simulation in a crystalline environment (L. Mollica et al., J. Phys. Chem. Lett. 3, 3657–3662 (2012)). The left panel presents a simplified representation of the link between small- and large-amplitude backbone motional modes. At low temperatures, the protein backbone is constrained to small amplitude modes separated by low-energy barriers, within substates separated by high barriers. As temperature increases, these modes are dynamically sampled, enabling larger-amplitude anisotropic modes. From Lewandowski, J.R., Halse, M.E., Blackledge, M. and Emsley, L. (2015) *Direct observation of hierarchical protein dynamics* Science, 348, 578–581. Reprinted with permission from AAAS.

accordingly or truncate (i.e., remove) the side chain at that place where electron density is no longer visible. A complementary approach would be to measure the mass of the pure, final protein before crystallisation to confirm if it is present or not. Indeed, mass spectrometry has grown impressively in its power and scope these last decades (see e.g. Liko et al. 2016).

A milestone example of finding out the damage of excessive X-ray dose to acquire crystallography diffraction data was when X-ray absorption spectroscopy was used with a much lower dose to study the reduction, by the X-ray-generated electrons, of the manganese ions in the oxygen evolving complex, the catalytic centre, of the photosystem II (Yano et al. 2005). Specific X-ray damage to protein disulphides as an effect of X-ray dose was shown much earlier, following a suggestion by Greg Petsko, Helliwell, and coworkers (1988), and is now an extensive field of enquiry as measured, for instance, by the succession of radiation damage conferences published in issues of the *Journal of Synchrotron Radiation.*

Infrared spectroscopy offers a variety of possible measurements to provide accuracy based on a precise crystal structure (see e.g. Barth 2007). It can be applied to both static and time-resolved studies. These latter include taking the difference spectra between two structural states. As with colour changes, which feature pivotally in two of my case studies below, such difference spectra can be very informative and can confirm and complement what is seen with 3D structural probes.

Solution small-angle X-ray scattering (SAXS) is used as a complement to cryocrystallography with a double aspect, confirming that the solution and solid-state crystal structures of a protein are the same and, as well, that the cryo and room temperature structures are the same. SAXS and its neutron equivalent small-angle neutron scattering have very wide-ranging applications and are extensively described in the recent books Svergun et al. 2013 and Lattman et al. (2018).

The connection between the molecular level of detail and the cellular level is bridged by the microscopies. The thickness of individual cells is problematic for electrons but not for X-rays. These offer resolution beyond the capability of light microscopy. A comparison of the capabilities of electron and X-ray microscopies is described in Du and Jacobsen (2018).

To close off this section, I choose an example from before protein crystallography that had been brought to fruition by Max Perutz and John Kendrew working in Cambridge. This is the case of sickle cell anemia and haemoglobin (Pauling et al. 1949). In the opening of the discussion section of their article Pauling et al. (1949) stated "*On the Nature of the Difference between Sickle Cell Anemia Hemoglobin and Normal Hemoglobin: The most plausible hypothesis is that there is a difference in the number or kind of ionizable groups in the two hemoglobins.*" Such changes were later shown by Max Perutz to have occurred and had created sticky patches, causing two haemoglobin molecules to hold together, distorting an erythrocytic cell and impeding blood flow. This is explained as a "PDB 101" here https://pdb101.rcsb.org/motm/41 (see the section entitled Troubled Haemoglobins). This example shows the importance of 3D structure as well as the measurements made by Pauling et al. (1949).

REFERENCES

Almeida, J. D. & Tyrrell, D. A. J. (1967) The morphology of three previously uncharacterized human respiratory viruses that grow in organ culture. *J. Gen. Virol.* 1, 175–178.

Andrews, S. J., Papiz, M. Z., McMeeking, R., Blake, A. J., Lowe, B. M., Franklin, K. R., Helliwell, J. R. & Harding, M. M. (1988) Piperazine Silicate (EU-19) - The structure of a very small crystal determined with synchrotron radiation. *Acta Cryst. B* 44, 73–77.

Barth, A. (2007) Infrared spectroscopy of proteins. *Biochim. Biophys. Acta* 1767, 1073–1101.

Bragg, W. L. (1968) X-ray crystallography. *Sci. Am.* 219, 58–70.

Bryce, D. L. (2017) NMR crystallography: Structure and properties of materials from solid-state nuclear magnetic resonance observables. *IUCrJ* 4, 350–359.

Du, M. & Jacobsen, C. (2018) Relative merits and limiting factors for x-ray and electron microscopy of thick, hydrated organic materials. *Ultramicroscopy* 184, 293–309.

Gemmi, M., Mugnaioli, E., Gorelik, T. E., Kolb, U., Palatinus, L., Boullay, P., Hovmöller, S. & Abrahams, J. P. (2019) 3D electron diffraction: The nanocrystallography revolution. *ACS Central Sci.* 5, 1315–1329.

Harris, R. K., Wasylishen, R. E. & Duer, M. J. (2009) *NMR Crystallography.* John Wiley, Chichester.

Helliwell, J. R. (2020) DeepMind and CASP14. *IUCr Newsletter* 28(4), 6.

Helliwell, J. R. (2021) Combining X-rays, neutrons and electrons, and NMR, for precision and accuracy in structure–function studies. *Acta Cryst. A* 77, 173–185.

Helliwell, J. R. (2022) Lecture at CCP4 2022 for which a recording is available here. https://www.youtube.com/watch?v=I7e2EYrmBMg&list=PLrmG39_bWIGgvTugFLZaxHkR0WCkf8hQQ&index=24.

Helliwell, M., Helliwell, J. R., Kaucic, V., Zabukovec Logar, N., Teat, S. J., Warren, J. E. & Dodson, E. J. (2010) Determination of zinc incorporation in the Zn substituted gallophosphate ZnULM-5 by multiple wavelength anomalous dispersion techniques. *Acta Cryst. B* 66, 345–357.

Henderson, R. (1995) The potential and limitations of neutrons, electrons and X-rays for atomic resolution microscopy of unstained biological molecules. *Q. Rev. Biophys.* 28, 171–193.

Kühlbrandt, W. (2014) The resolution revolution. *Science* 343, 1443–1444.

Lattman, E. E., Grant, T. D. & Snell, E. H. (2018) *Biological Small Angle Scattering an IUCr Monograph.* Oxford University Press, Oxford.

Lewandowski, J. R., Halse, M. E., Blackledge, M. & Emsley, L. (2015) Direct observation of hierarchical protein dynamics. *Science* 348, 578–581.

Liko, I., Allison, T. M., Hopper, J. T. S. & Robinson, C. V. (2016) Mass spectrometry guided structural biology. *Curr. Opin. Struct. Biol.* 40, 136–144.

Martin, R. W. & Zilm, K. W. (2003) Preparation of protein nanocrystals and their characterization by solid state NMR. *J Magn. Reson.* 165, 162–174.

Middleton, D. A. (2009) Structural biology. In Harris, R. K., Wasylishen, R. E. & Duer, M. J. (Eds.), *NMR Crystallography*, Ch. 27, 417-433. John Wiley, Chichester.

Mollica, L., Baias, M., Lewandowski, J. R., Wylie, B. J., Sperling, L. J., Rienstra, C. M., Emsley[N1] , L.& Blackledge, M. (2012) Atomic-resolution structural dynamics in crystalline proteins from NMR and molecular simulation. *J. Phys. Chem. Lett.* 3, 3657–3662.

Neutze, R., Wouts, R., van der Spoel, D., Weckert, E. & Hajdu, J. (2000) Potential for biomolecular imaging with femtosecond X-ray pulses. *Nature* 406, 752–757.

Pauling, L., Itano, H. A., Singer, S. J. & Wells, I. C. (1949) Sickle cell anaemia a molecular disease. *Science* 110, 543–548.

Sanjuan-Szklarz, W. F., Woińska, M., Domagała, S., Dominiak, P. M., Grabowsky, S., Jayatilaka, D., Gutmann, M. & Woźniak, K. (2020) On the accuracy and precision of X-ray and neutron diffraction results as a function of resolution and the electron density model. *IUCrJ* 7, 920–933.

Schwarzenbach, D., Abrahams, S. C., Flack, H. D., Prince, E. & Wilson, A. J. C. (1995) Statistical descriptors in crystallography: Report of the IUCr subcommittee on statistical descriptors. *Acta Cryst. A* 51, 565–569.

Svergun, D. I., Koch, M. H. J., Timmins. P. A. & May, R. P. (2013) *Small Angle X-ray and Neutron Scattering from Solutions of Biological Macromolecules*, an IUCr Text. Oxford University Press,Oxford.

Vainshtein, B. (1964) *Structure Analysis by Electron Diffraction*. Pergamon Press, Oxford.

Ward, A. B., Sali, A. J. & Wilson, I. A. (2013) Integrative structural biology. *Science* 339, 913–915.

Wüthrich, K. & Wagner, G. (1975) Proton NMR studies of the aromatic residues in the basic pancreatic trypsin inhibitor (BPTI). *FEBS Lett.* 50, 265–268.

Yano, J., Kern, J., Irrgang, K.-D., Latimer, M. J., Bergmann, U., Glatzel, P., Pushkar, Y., Biesiadka, J., Loll, L., Sauer, K., Messinger, J., Zouni, A. & Yachandra, V. K. (2005) X-ray damage to the Mn_4Ca complex in single crystals of photosystem II: A case study for metalloprotein crystallography. *Proc. Natl Acad. Sci. USA* 102, 12047–12052.

CHAPTER 4

Real-World Science Study Examples

BIOLOGICAL AND BIOMEDICAL SCIENCE: REDUCTIONISM VERSUS WHOLE ORGANISM BIOLOGY

Overall, there is scepticism about what we "atomic level structuralists," my area of expertise, do for biology:

Dame Ottoline Leyser, Professor of Plant Development at the University of Cambridge, was quoted recently (Turney 2019) that: "*The defining feature of biology during the past few decades has been figuring out details of the parts. But biological systems don't think they have parts.*" Scepticism of the role of reductionism in understanding biology is a theme in Ernst Mayr's book *What Makes Biology Unique?* (Mayr 2007). Mayr even argues against the relevance of the discovery of the DNA double helix to understanding biology. As a counterpoint, a founding father of quantum mechanics, the physicist Erwin Schrodinger (1943), posed the question "what is life?" in his influential book *What is Life?: The Physical Aspect of the Living Cell*, in effect applying the physical sciences to this central question of biology.

Then Sir Paul Nurse, awarded the Nobel Prize in Physiology or Medicine in 2001 for his work on the cell cycle of fission yeast, in an interview by Ireland (2014) quoted here:

DOI: 10.1201/9781003405399-4

> Rather than get bogged down in detailed molecular descriptions of everything, I'm asking bigger questions like "how does a cell know how big it is?," which has always fascinated me.

So, where do I think these strong views stand today? i.e. against our reductionist research in the molecular sciences, within which crystallography is a key player, as are microscopies and spectroscopies. More to the point, how can the crystallographer respond constructively, maybe only partly, to the concerns of the holistic biologists?

The issue of relevance starts when considering the protein crystalline state or the solution state of a protein but imagined placed inside the biological cell. I have already made comparisons of methods above in Chapter 3. Nuclear magnetic resonance (NMR) provides atomically detailed results in solution, and of course, protein crystallography provides atomic details for a protein in the solid state. The protein crystal is a curious hybrid of the solid state though being an ordered lattice but also with a large percentage of the crystal volume being solution, namely the solvent channels that run through the crystal.

Studies of the structure and function of an enzyme in the crystalline state, were to my mind, greatly facilitated by the invention of the flow cell (Wyckoff et al. 1967). Its use, flowing an enzyme's substrate to it via a pipe, answered with a resounding yes that these crystal structure results are relevant to enzyme catalytic function. Flow cell results overcome the objections of the NMR solution state spectroscopists to the crystallographer's results in the protein crystal.

Weaknesses in the armoury of crystallography remain, such as crystallisation conditions, to a greater or lesser degree, taking one's results away from biological functioning conditions. As Yibin Lin's article (Lin 2018) states:

> Structure based drug design requires accurate structural information with the native conformation of a protein. It is a big challenge to find crystallization conditions that are close to the physiological conditions of a protein.

This caveat of Yibin Lin (2018) is perhaps more relevant than the one about having a lattice for a crystal. Lattice contacts in macromolecular crystals are rather few, whereas other possible effects of non-physiological crystallisation conditions cannot be similarly dismissed as they affect all the atoms in the crystallized protein.

A similar concern is the question about the strict relevance to biology of crystallography results now predominantly based on X-ray diffraction data measured at cryo temperature (Halle 2004). Conducting crystallography at physiologically relevant temperatures has then become an objective.

Neutron macromolecular crystallography (nMX), whilst pursuing protein structures with protonation states experimentally determinable, has also automatically yielded room-temperature structures. Given also that projects succeed in getting neutron beamtime only where all other methods have failed, X-ray, electron, or NMR-based, in structural biology there is a strategic importance of the nMX method. Neutron macromolecular crystallography has seen a sustained growth of the instruments, the software, and methods at leading neutron sources (e.g. see Blakeley 2009).

The new femtosecond-range X-ray lasers yield X-ray diffraction data at room temperature, and before radiation damage can kick in, the "diffract before the sample is destroyed" approach (Neutze et al. 2000). Synchrotron facilities are now also adopting X-ray laser methods for delivery of streams of micron-sized samples and thereby yielding results at physiological temperatures albeit not free of radiation damage like the X-ray lasers.

So, the physical methods of crystallography, microscopy, and spectroscopy continue to strive for and clearly deliver functionally relevant structural results. While many biologists have embraced molecular structure, shall we molecular-level biologists ever convince holistic biologists? Our studies rarely take us directly to the whole organism.

I can give a case study, though, where the most important bridge to whole animals is achieved: the colouration of the lobster shell. Spanning more than 50 years, the methods of biochemistry, biological crystallography, spectroscopy, solution X-ray scattering, and microscopy have been applied to study the molecular basis of the colouration of the live lobster. The blue colouration of the carapace of the European and American lobsters is provided by a multimolecular carotenoprotein, α-crustacyanin. The European lobster (*Homarus gammarus*) crustacyanin has been the most extensively studied. Its α-crustacyanin complex, extracted from the lobster carapace, is a 16-mer of different protein subunits each binding the carotenoid, astaxanthin. The biological purpose of the coloured shell is unknown although it may be a means of camouflage against predators such as octopus. A breakthrough in the structural studies came from the determination of the crystal structure of β-crustacyanin, comprising the protein subunits A1 with A3 with two shared-bound astaxanthins (Cianci et al. 2002). This crystal structure was based on diffraction data measured at 100K. The crystal was blue at room temperature and blue at cryo temperature from which we can conclude that the 3D structure of the β-crustacyanin is free of cryoartefacts, at least in terms of the protein astaxanthin to protein molecular interactions. Furthermore, solution X-ray scattering (SAXS) measurements of the β-crustacyanin showed an excellent agreement with the predicted SAXS curve from the cryo crystal structure (Chayen et al. 2003). So, we can conclude that the crystalline state and the solution state structures of β-crustacyanin agree. Colour is a simple but effective assay for a real-life connection whereby the mechanism of change in colouration on cooking a lobster, to orange red, can be explicitly seen, and compared with crystal colours.

PHYSICS, CHEMISTRY, AND MATERIALS: CASE STUDY OF BATTERIES

Ever since the discovery of “the battery” by Volta in Italy, there has been intense interest in their evolution, leading to improvements in portability and thereby the widening of their scope. From the outset, Napoleon Buonaparte invited Volta to demonstrate the first battery to him and his generals (see Figure 4.1).

In 2019, the Nobel Prize in Chemistry was awarded jointly to John B. Goodenough, M. Stanley Whittingham, and Akira Yoshino "for the development of lithium-ion batteries". In their press release (https://www.nobelprize.org/prizes/chemistry/2019/press-release/), the Nobel Prize Chemistry Committee summarized why the award was made because of the:

> “lithium-ion battery. This lightweight, rechargeable, and powerful battery is now used in everything from mobile phones to laptops and electric vehicles. It can also store significant amounts of energy from solar and wind power, making possible a fossil fuel-free society.”

This Nobel Prize 2019 website has splendid schematics of each of the working batteries of the three Nobellists neatly showing the evolution of their ideas and experimental battery designs.

FIGURE 4.1 Volta demonstrating his battery pile to Napoleon. I came upon this picture during my visit to the Volta Museum in Como, Italy; The origin of the picture is unknown.

A key research paper from the Goodenough laboratory was by Mizushima et al. (1980), whose laboratory electrochemical cell is shown in Figure 4.2.

A key feature of the characterisation of the electrochemical cell was to assess the internal structure using X-ray powder diffraction as the lithium-ion fraction was varied, which allowed the authors to conclude that there is:

> "ordering of lithium and cobalt into alternate (III) layers; not true of the corresponding nickel compounds." *Furthermore that* "The data's agreement between observed and calculated X-ray powder diffraction data imply a retention of the layer structure on removal of lithium atoms."

The reversible insertion and removal of lithium atoms in the crystalline layers of the cobalt oxide was a pivotal aspect of the operation of this new type of battery material.

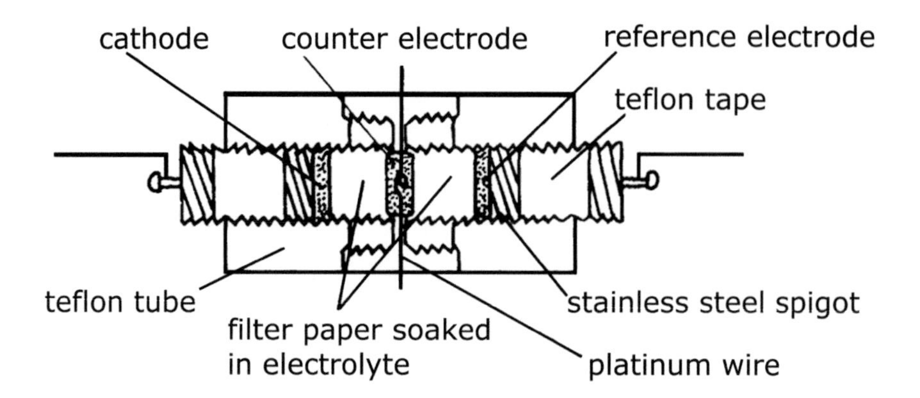

FIGURE 4.2 Cell for the electrochemical measurements. Reproduced from Mater. Res. Bull. Volume 15, pages 783–789 authors Mizushima, K.; Jones, P. C.; Wiseman, P. J.; Goodenough, J. B. (1980), article entitled "*Li_xCoO_2 (0 < x < 1): A New Cathode Material for Batteries of High Energy Density.*" with the permission of Elsevier.

In terms of showing scientific truth to the public and to schoolchildren, since so much of what we do in science is quite abstract, a battery is something we all use and benefit from, and thereby a working example of scientific truth. Science delivers and, in the battery, it is an example to rival any science example documenting the truth.

TRUTH IN DISPUTE IN THE COURTROOM: PHARMACEUTICAL POLYMORPHS AND COMPANY AGAINST COMPANY PATENT DISPUTES

The culture of science and its seeking of truth has an interesting counterpart in the courtroom, where, depending on the country, innocence is presumed to be the case for an accused person until the truth establishes otherwise. To proceed to a conclusion, there is a burden of proof, which is required of the party making the accusation. This requires the production of evidence to establish the truth of the facts. The more serious the factual matter is, the higher the standard of proof that will be required. In 'less serious cases' the standard of proof required allows for a "balance of probabilities." Much loved moments in a courtroom TV drama are when one by one witnesses take the stand and each must swear an oath "*to tell the truth, the whole truth, and nothing but the truth.*" This became the title of my book, of course.

In what types of cases has the legal culture met the scientific one? Well, first of all, science finds its way into criminal cases via forensic evidence found at the scene of a crime, which can include the DNA profile of the person accused of committing the crime. These pieces of evidence are usually so definitive as to be unarguable because that evidence clearly establishes the truth.

A more disputatious situation can arise when two pharmaceutical companies argue in the courtroom about the legitimacy of a new patent based on the discovery by one of them of a new polymorph of a drug compound. Billions of pounds, dollars, or euros can be at stake. There are many well-documented examples of such cases (Bernstein (2020) chapter 10).

An interesting case described by Sherwood et al. (2022) is that of psilocybin {systematic name: 3-[2-(dimethylamino)ethyl]-1H-indol-4-yl dihydrogen phosphate} which is a zwitterionic tryptamine natural product found in numerous species of fungi known for their psychoactive properties (see Figure 4.3).

A zwitterion is a molecule that contains an equal number of positively and negatively charged functional groups. Following psilocybin's original structural elucidation and chemical synthesis in 1959, purified synthetic psilocybin was evaluated in clinical trials and showed promise in the treatment of various mental health disorders. This compound became the subject of a patent legal dispute between the Usona Institute, in which two of the authors of Sherwood et al. (2022) worked. A non-profit organization, Freedom to Operate, used research from chemists and crystallographers to argue in a legal filing that a company called Compass Pathways Limited's claims that a form of synthetic psilocybin they had found was not an invention (Londesbrough et al. 2019). This research by Sherwood et

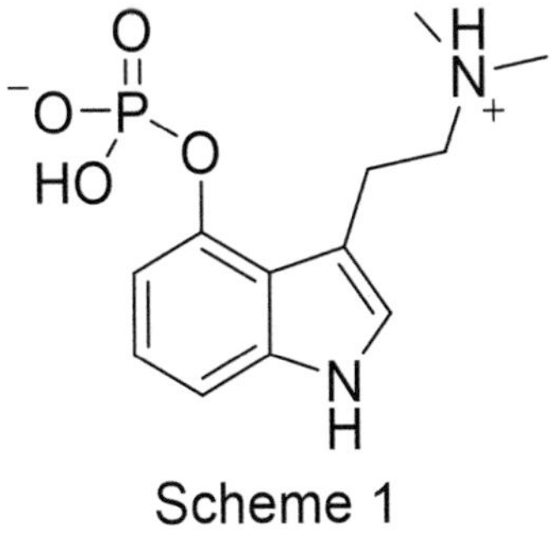

Scheme 1

FIGURE 4.3 The psilocybin molecule. Reproduced with the permission of the corresponding author (Dr James Kaduk) of Sherwood et al. 2022 and the IUCr Journals.

al. (2022) firmly established that there are three crystalline forms of psilocybin, namely: "Hydrate A, Polymorph A, and Polymorph B" (see Figures 4.4, 4.5 and 4.6).

There is therefore no Polymorph A', as claimed by Compass. (Sherwood et al. 2022) use the designation Polymorph A' to distinguish it from their Polymorph A; the patent names it Polymorph A). To challenge a patent required meticulous proof, and that is what Sherwood et al. (2022) provided. Another dimension to this legal dispute is that Compass, unlike Usona, is a for-profit company that went public in September 2020.

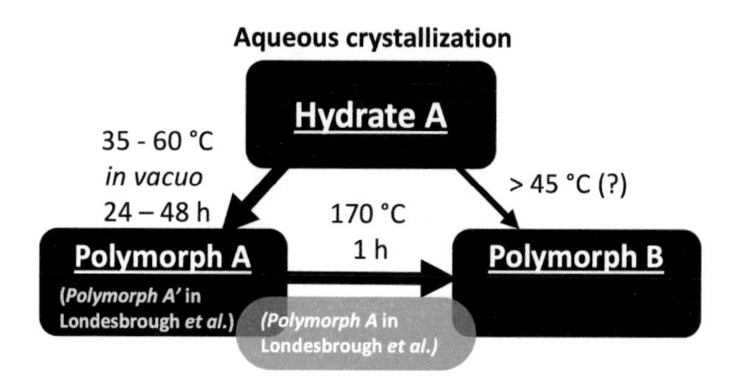

FIGURE 4.4 This shows the preparation pathways for the solid-state forms of psilocybin. Reproduced with the permission of the corresponding author (Dr James Kaduk) of Sherwood et al. 2022 and the IUCr Journals.

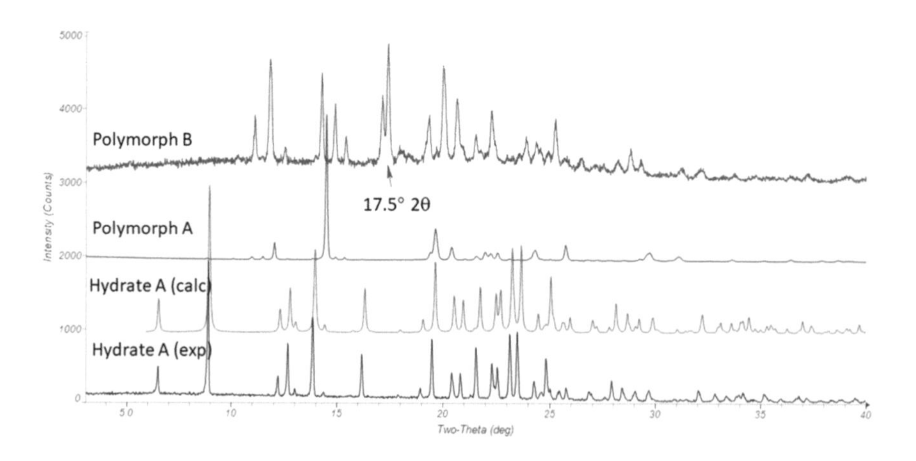

FIGURE 4.5 This shows the powder X-ray diffraction patterns for polymorphs A and B as well as the hydrated form. Reproduced with the permission of the corresponding author (Dr James Kaduk) of Sherwood et al. 2022 and the IUCr Journals.

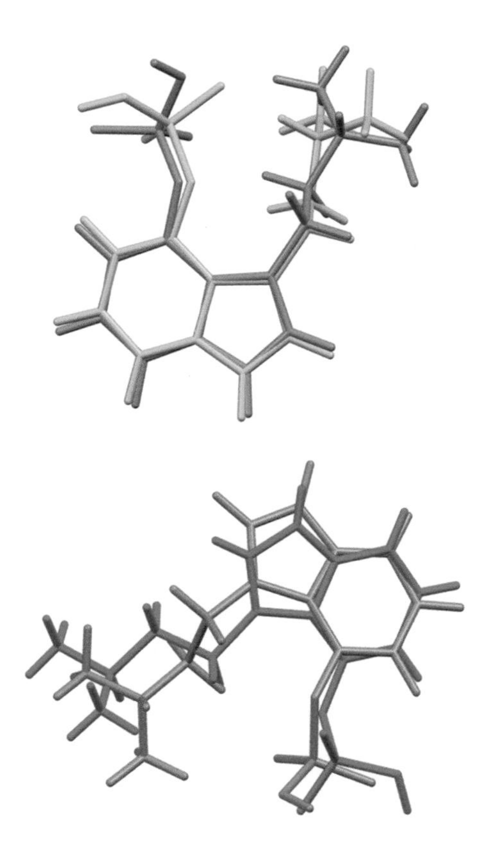

FIGURE 4.6 This shows the molecular structures of Top: the two Polymorphs A and B and Bottom: the hydrate and Polymorph A. Reproduced with the permission of the corresponding author (Dr James Kaduk) of Sherwood et al. 2022 and the IUCr Journals.

What is a patent? Patents have the intended purpose to reward innovation by giving an inventor the right to exclude others from benefiting from their invention for typically 20 years. The biomedical and pharmaceutical worlds regularly exercise this right when developing new medicines. Patents are granted on inventions when they show that they are new, not obvious to someone else with expertise in their field, and that they are useful.

What are polymorphs? These are phases in which molecules arrange themselves in a crystal form in different ways, and those different ways are called polymorphs. Why are these of such commercial interest? Different crystal forms, the polymorphs, can have varying properties, like melting points or solubility. This has relevance in pharmaceuticals—one crystal form, for example, can be easier to make into a consumer product than another or work more effectively in the patient.

The evidence that Compass used and cited in their patent for their new form of psilocybin was that:

> Hydrate A is the only polymorphic form existing across a range of temperatures with no diffraction peak in the 17 degrees 29 minutes region…that Polymorph A is the true form with Polymorph A' formed at a small scale being atypical.

This evidence is what formed the core of the dispute as Usona, and its lawyers argued that it had misdescribed the forms of this molecule. They based their evidence on that of Sherwood et al. (2022). In the petition by Usona and its lawyers, world-renowned powder diffraction expert Dr. James Kaduk, a coauthor of Sherwood et al. (2022), described how he determined the psilocybin crystal structures from the powder X-ray diffraction data for the three well-known polymorphs and how he then mathematically modelled mixtures of those different polymorphs to predict what the corresponding X-ray diffractograms would look like. The measurements featured samples of synthetic psilocybin from stocks held over many decades. By analysing such samples, it allowed Dr. Kaduk to test the hypothesis that it has been these three forms of crystalline psilocybin (trihydrate, A and B) that have been produced again and again, in varying combinations, and that "Polymorph A prime" was not a new polymorph. The Sherwood et al. (2022) conclusions were:

> In this article, we show conclusively that all published data can be explained in terms of three well-defined forms of psilocybin and that no additional forms are needed to explain the diffraction patterns.

The psilocybin case is an example from the USA, where a Patent Trial and Appeal Board forms a collection of administrative judges with expertise in science and patent law to decide whether a petitioner demonstrates that at least one of the challenged claims is "more likely than not" unpatentable—and whether the case should proceed to a trial. The psilocybin case was resolved in favour of Usona. Clearly, a pivotal factor in this dispute is that the study by Sherwood et al. (2022) was in a highly respected academic crystallography journal, where it was carefully peer-reviewed.

An extensive news magazine piece on the psilocybin case is by Love (2021).

REFERENCES

Bernstein, J. (2020) *Polymorphism in Molecular Crystals*. 2nd edition. OUP, Oxford. Chapter 10 "Polymorphism and patents".

Blakeley, M. P. (2009) Neutron macromolecular crystallography. *Crystallogr. Rev.* 15(3), 157–218.

Chayen, N. E., Cianci, M., Grossmann, J. G., Habash, J., Helliwell, J. R., Nneji, G. A., Raftery, J., Rizkallah, P. J. & Zagalsky, P. F. (2003) Unravelling the structural chemistry of the colouration mechanism in lobster shell. *Acta Cryst. D* 59, 2072–2082.

Cianci, M., Rizkallah, P. J., Olczak, A., Raftery, J., Chayen, N. E., Zagalsky, P. F. & Helliwell, J. R. (2002) The molecular basis of the coloration mechanism in lobster shell: β-crustacyanin at 3.2 Å resolution. *PNAS USA* 99, 9795–9800.

Halle, B. (2004) Biomolecular cryocrystallography: Structural changes during flash-cooling. *Proc. Natl Acad. Sci. USA* 101, 4793–4798.

Ireland, T. (2014) Crick questions: Sir Paul Nurse. *Biologist* 61, 32–35.

Lin, Y. (2018) What's happened over the last five years with high-throughput protein crystallization screening? *Exp. Opin. Drug. Discov.* 13, 691–695.

Londesbrough, D. J., Brown, C., Northen, J. S., Moore, G., Patil, H. K. & Nichols, D. E. (2019). *US Patent 10519175B2*. Compass Pathways Ltd.

Love, S. (2022) *Inside the Dispute Over a High-Profile Psychedelic Study.*Vice Magazine. https://www.vice.com/en/article/4awj3n/inside-the-dispute-over-a-high-profile-psychedelic-study.

Mayr, E. (2007) *What Makes Biology Unique?* Cambridge University Press, Cambridge, UK.

Mizushima, K., Jones, P. C., Wiseman, P. J., Goodenough, J. B. (1980) $Li_xC_oO_2$ ($0 < x < 1$): A new cathode material for batteries of high energy density. *Mater. Res. Bull.* 15, 783–789.

Neutze, R., Wouts, R., van der Spoel, D., Weckert, E. & Hajdu, J. (2000) Potential for biomolecular imaging with femtosecond X-ray pulses. *Nature*, 406, 752–757.

Schrödinger, E. (1943) What is life? The physical aspect of the living cell, based on lectures delivered under the auspices of the Dublin Institute for Advanced Studies at Trinity College, Dublin, in February. Reprinted by Cambridge University Press, 2012.

Sherwood, A. M., Kargbo, R. B., Kaylo, K. W., Cozzi, N. V., Meisenheimer, P. & Kaduk, J. A. (2022). Psilocybin: Crystal structure solutions enable phase analysis of prior art and recently patented examples. *Acta Cryst. C* 78, 36–55.

Turney, J. (2019) The puzzle of life. *Times Higher Educational Suppl.* 21st February, 44–45.

Wyckoff, H. W., Doscher, M., Tsernoglou, D., Inagami, T., Johnson, L. N., Hardman, K. D., Allewell, N. M., Kelly, D. M. & Richards, F. M. (1967) Design of a diffractometer and flow cell system for X-ray analysis of crystalline proteins with applications to the crystal chemistry of ribonuclease-S. *J. Mol. Biol.* 27, 563–578.

CHAPTER 5

Don't Take My Word for It

FAIR and FACTual Data

THE BASIS OF TRUST requires:

- Trust in the scientist: their articles, their database entries, and their data sets, including data quality and data set completeness.
- Trust in the processes of science: in turn, this requires careful recording of metadata for the experiment, the processing, and the deriving of the fitted model workflows. These form the provenance.
- Trust in the measurement at source: the raw data as the ground truth of a science study.

In an education article that I wrote with a colleague within one of our community journals (the Journal of Applied Crystallography; Helliwell and Massera 2022), both of us crystallographers, we included Figure 5.1 to show up the central position of trust in science and how it is built up from different facets, i.e., the words that surround the word "trust."

DOI: 10.1201/9781003405399-5

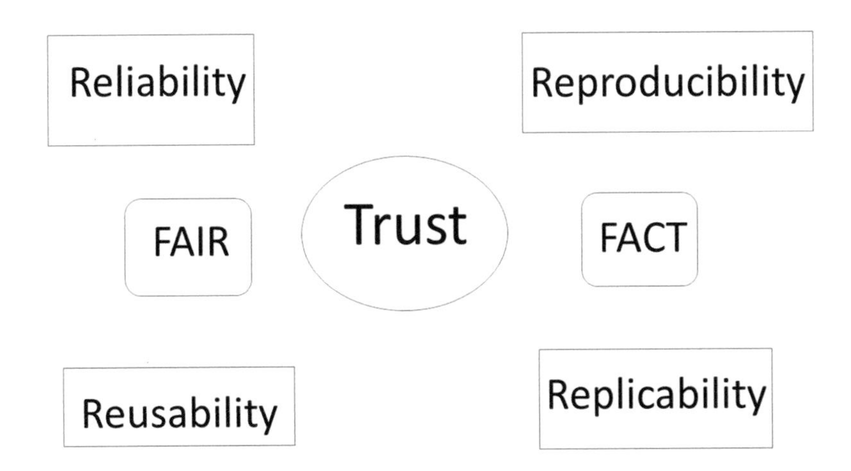

FIGURE 5.1 Trust in science is built up from different facets. Is the study reliable, if so, how is reliability assessed? The reader of a publication will want access to, and thereby be able to reuse, the underpinning data (and the software used) in a study to try to reproduce the calculations. Other authors will maybe attempt their own fresh study to try and replicate the first study. The terms FAIR and FACT are described in the main text. This figure was first presented by Helliwell and Massera (2022). Reproduced with the permission of IUCr Journals.

Trust in science is generally assumed by scientists and is yet ever more under scrutiny if there are failures. From the article *Trust in Science* (Barber 1987):

> Trust is an essential constituent of all social relationships and all societies.
>
> One sense of trust refers to an expectation or prediction that an assigned or accepted task will be competently performed.

Also, scientists very much expect that a qualified scientist can be trusted.

These two forms of trust are quite distinct from each other. (Or is it faith? We certainly tend to believe our predecessors and not feel that we must reproduce all previous results).

So, the apprentice scientist must learn to be trustworthy in both senses.

The wider science data scene is celebrating the FAIR data accord, namely, that data be Findable, Accessible, Interoperable, and Reusable [Wilkinson et al., "Comment: The FAIR guiding principles for scientific data management and stewardship," Sci. Data 3, 160018 (2016)]. Some social scientists also emphasize more than FAIR being needed, the data should be "FACT," which is an acronym meaning Fairness, Accurate, Confidential, and Transparent [van der Aalst W. M. P., Bichler M., and Heinzl A., "Responsible data science". Bus Inf. Syst. Eng. 59(5), 311–313 (2017)], this being the issue of ensuring reproducibility not just reusability.

Whereas FAIR looks at practical issues related to the sharing and distribution of data (Wilkinson 2016), FACT focuses more on the foundational scientific challenges (van der Aalst et al. 2017). In crystallography, the requirement for FAIR data is satisfied by our databases for processed diffraction data and their derived molecular models.

Van der Aalst et al.'s (2017) perception of trust, by the public, is very good I think:

> Data science can only be effective if people trust the results and are able to correctly interpret the outcomes. The journey from raw data to meaningful inferences involves multiple steps and actors, thus accountability and comprehensibility are essential for transparency.

An influential report was published in 2015 (Science International 2015) "Open Data in a Big Data World". In a general way, this landmark document described "*the values of open data in the emerging scientific culture of big data.*" The International Union of Crystallography (IUCr) acknowledged the importance of this accord and endorsed the analysis of the values of open data and the principles of open data set out in the document. Because the specific values, significance, and implementation of Open Data principles vary in detail between disciplines, the IUCr considered it useful to contribute a detailed response to the accord, as a case study of best practice emerging in one field, that of crystallography, and which I coauthored (Hackert et al. 2016). We tabled this document at the International Data Week held in Denver in 2016. Two key elements of our paper were that scientific data should be:

- ***complete*** *(i.e., all data collected for a particular purpose should be available for subsequent re-use); and*
- ***precise*** *(the meaning of each datum is fully defined, processing parameters are fully specified and quantified, and statistical uncertainties evaluated and declared).*

We also emphasized that, in practice and over the history of sustaining efforts in crystallography, there is a diverse ecosystem of disciplinary databases, data repositories, experimental facilities, and publishers, some sustained through subscription-based access but at no charge to the author/depositor, and without public funding. At the present time, these varieties of approaches to sustainability and quality assurance serve this discipline well. The full linking of article and data is another key element of openness and of "Don't take my word for it" the title of course of this Chapter 5.

Hackert et al.'s case study (2016) was well received at the International Data Week 2016 and it has remained as the sole discipline specific detailed response to the Science International (2015) document.

DATA OPPORTUNITIES TODAY: DATABASES AND THEIR ENSURING ESTIMATES OF A CRYSTAL STRUCTURE'S RELIABILITY

In crystallography, a quality indicator of a crystal structure from the experimental measurements is characterized by an R factor or residual between the measured data and the calculated data from the model. In fact, in protein crystallography, two R factors are calculated, where one of them is known as the Rfree and is based on typically 5% of the diffraction data randomly distributed through the measurements and excluded from the refinement of the model. Thereby, 95% of the measurements guide the updating of the model, i.e., its optimisation and if the calculations are on the right track, then the Rfree will also improve. R is also commonly referred to as the reliability of the crystal structure model. A perfect model would, of course, have an R value of zero.

To improve the processes of arriving at truth, the crystallography community uses the tool of the inter laboratory round robin; see eg Helliwell et al. 1981 to improve diffraction data processing software. A second approach is to use experimental crystal structure database entries for the same compound from different laboratories; an example of this is Taylor and Kennard (1986), who compared 100 such pairs. These pairs of experimental studies may well be under what are imagined as identical conditions. The compound is provided pure by a chemicals' supply company in separate bottles to the two different crystallography laboratories, which are equipped with X-ray diffractometers, sometimes even purchased from the same manufacturer. Then the software used for the data processing and the molecular model refinement may be different or the same. So, do the paired studies agree? Also, does the precision calculated for each member of the pair agree? Taylor and Kennard (1986) found that each study's estimated standard deviations, as they were called at that time (these words later changed to standard uncertainties, with abbreviation s.u.), were about 40% lower in the individual studies than the values calculated by analysis of the agreement of the pairs. They found one pair where the crystal structures were not comparable at all. They contacted the two laboratories and found that the crystal sample studied at each place were systematically different.

RAW DATA ARCHIVES SUCH AS ZENODO AND GLOBAL OPEN SCIENCE CLOUD

Scientific publication with data, reaches towards objectivity and weeds out falsification cases. If we can preserve the raw data as well as the processed data and the final derived model, along with the metadata (these are the data describing the data) at each stage, then we have a complete workflow of the analysis. Raw data are the most challenging in terms of file sizes, but as digital archives have expanded in the past decade or so, it is increasingly possible to preserve these data too and can be termed the "ground truth." See Figure 5.2.

Rather provocatively perhaps, I have labelled a direction of travel of the research from objectivity to subjectivity in Figure 5.2. What do I mean by that? The objective stage is possible but requires archiving of the raw data, which is the most burdensome in terms of sizes of files. During processing of the raw diffraction data, decisions are made by the researcher and subjectivity is introduced. During model refinement, further decisions are made and more subjectivity is introduced in arriving at the final protein "model." Figure 5.3 shows examples of these stages of data in crystallography.

The article narrative describes the methods and decisions made by authors. The interpretation introduces a further level of subjectivity and may even include a wish to see a hypothesis "proven." A publication is an important narrative of the work done and interpretations made by researchers securing a scientific discovery. The Royal Society's motto (https://royalsociety.org/about-us/history/) neatly states, "Nullius in verba" ("Take nobody's word for it"), whereby the role of the underpinning facts is paramount. The raw experimental data are a key component of those facts. Thereby, the objectivity that preserving those raw data with the article gives is because readers can check each decision of the authors.

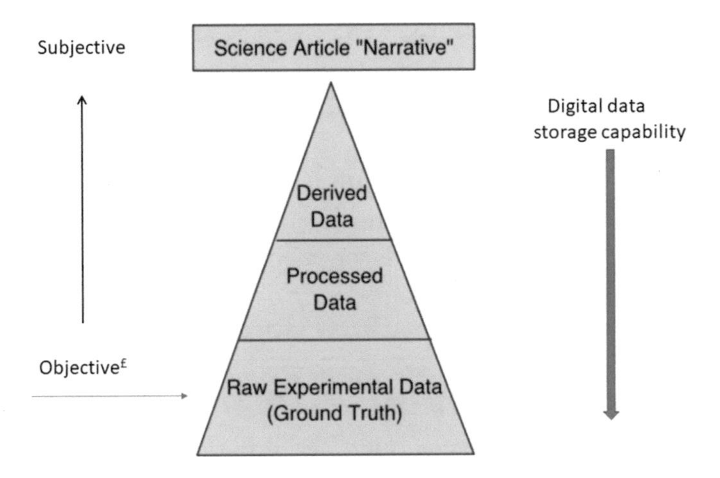

FIGURE 5.2 Is objectivity in science possible? This is discussed in more detail in Helliwell (2019). The £ superscript means that there is a caveat that an apparatus must be calibrated by a person and so still leaves a trace of subjectivity; nevertheless the raw data are as close to objectivity and ground truth that we can get.

Previous articles on this theme that I have written with several colleagues have addressed the how and what, e.g., describing the archiving of raw diffraction data sets (Tanley et al. 2013), and the metadata that are essential to be included (Kroon-Batenburg and Helliwell (2014),

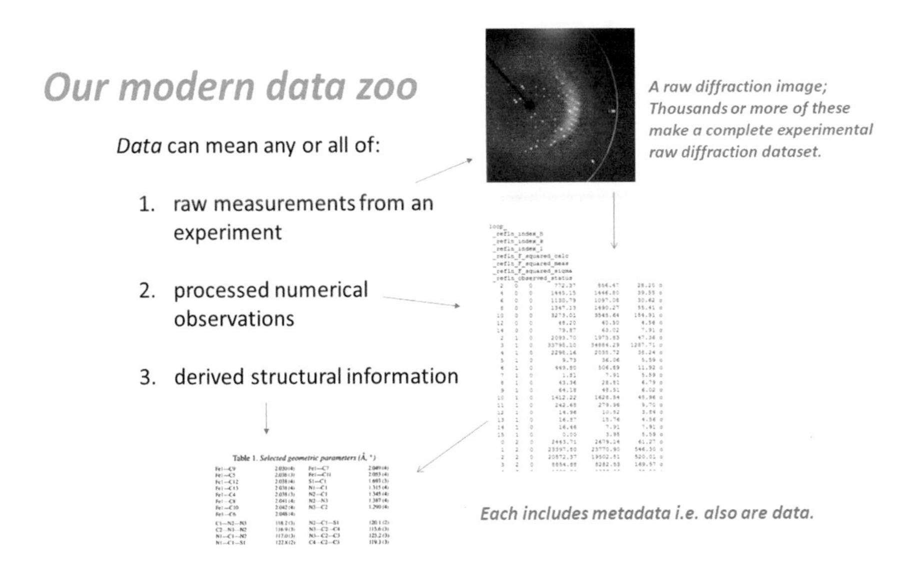

FIGURE 5.3 Examples of the stages of raw, processed, and derived data in crystallography.

Kroon-Batenburg et al. (2017)) as well as ***why*** we should archive raw diffraction data (Helliwell et al. 2017). Crystal structure analysis as a discipline is widely recognized as achieving FACT and FAIR with its data over many decades.

We can cite the following community awards and recognition. Firstly, the Association of Learned and Professional Society Publishers' Award was made in 2006 to IUCr Journals for publication innovation regarding linking of articles with their underpinning data. Secondly, the CODATA Prize was awarded in 2014 to Professor Sydney R. Hall, *Editor of Acta Cryst Section C Crystal Structure Communications.* The award citation states, "*He devised a universal file format and Crystallographic Information Framework (CIF), a momentous contribution in the area of data characterisation and publications standard. It enables data validation for articles published by IUCr journals.*"

More generally, thirdly, the 2018 European Union Report "Turning FAIR Data into Reality" (European Commission 2018) stated:-

- *"The requirement from academic journals that authors provide data in support to their papers has proven to be potentially culture-changing, in crystallography;"*
- *"Many data standards are maintained by international scientific unions (e.g., the International Union of Crystallography)."*

The above are two quotes from https://ec.europa.eu/info/sites/info/files/turning_fair_into_reality_1.pdf

Further efforts of IUCr to ensure crystallographic data are FACT that involve leading by example with its own journals so that in Acta Cryst. B, C, E, and IUCrData the article narrative, the automatic "general" validation checks (checkcif) and the underpinning data are checked thoroughly by subject specialists (i.e., the specialist referees). Efforts are underway to extend this specialist data checking to biological crystallography (Helliwell 2019) and indeed all areas of crystallography and structural science publishing. Also, more recently, IUCr journals encourage and expedite citations of the dois for raw diffraction data sets publications; examples can be found in IUCrJ, Acta Cryst D, and Acta Cryst F. A checkcif for raw diffraction data has been developed by the IUCr Committee on Data jointly with the IUCr Committee for the Maintenance of the CIF Standard (COMCIFS) and the PaNOSC EU Project (https://www.panosc.eu/). Crystallography as a discipline is also striving for our data to be FACT and FAIR now also with Big Data Archives; thereby striving for ultimate objectivity. Indeed as Strickland et al. 2008 remarked: "*Ideally, the full scientific record should provide access to the raw data…the IUCr is beginning to consider longer-term approaches to archiving the raw data.*"

Across all the sciences, why is raw diffraction data archiving important? The answers are that: it allows reuse to reproduce or even extend the complete workflow of a data analysis, and it can serve as benchmarks in developing improved methods of analysis, software, and algorithms. In crystallography, in addition to the above, it allows checking the interpretation of the symmetries of the crystals; it facilitates analysing diffraction from multiple lattices present in the crystals. Also, the analysis of the diffuse scattering that reflects correlated motions or disorder of atoms in the crystals is possible.

Please note though that there is a qualification to my argument that archiving our raw diffraction data allows us to attain full objectivity. As explained more generally at the Stanford Encyclopaedia of Philosophy section on Scientific Objectivity https://plato.stanford.edu/entries/scientific-objectivity/ Section 2.3, which it calls "***The Experimenter's Regress*** (and I paraphrase)":

> reasonable calibration of the instrument takes us from objectivity to a level of subjectivity.

So, our detector calibrations, which we must judge as reasonable, are required for our raw diffraction data to be deemed acceptable.

Overall, there is a philosophical view of the importance of access to raw diffraction data, namely analysis through one's own eyes not the lens of someone else. For case studies, see Helliwell et al. (2017). The IUCr and its Committee on Data's take-home message is that the IUCr (representing the community of crystallographers and structural scientists) maintains the need for the highest quality of data management at all stages, from experimental data collection through reduction and analysis to publication and database deposition.

WORKING WITH THOSE ALREADY MEASURED DATA BY OTHER SCIENTISTS; LOOKING FOR PATTERNS: THE PROTEIN FOLDING PROBLEM WAS SOLVED THIS WAY

The Critical Assessment of Techniques for Protein Structure Prediction (CASP) challenge is a major and very well done effort these last decades carried out by global organizers: Dr. John Moult (chair), University of Maryland, USA; Dr. Krzysztof Fidelis, UC Davis, USA; Dr. Andriy Kryshtafovych, UC Davis, USA; Dr. Torsten Schwede, University of Basel and SIB Swiss Institute of Bioinformatics, Switzerland; and Dr. Maya Topf, Birkbeck, University of London, UK, and CSSB (HPI and UKE) Hamburg, Germany. The CASP14 Press release of 30 November 2020 https://predictioncenter.org/casp14/doc/CASP14_press_release.html stated that:

> Today (Monday), researchers at the 14th Community Wide Experiment on the CASP14 will announce that an artificial intelligence (AI) solution to the challenge has been found.by DeepMind's AlphaFold program.

This is clearly, i.e., all the expert judges of the CASP series of challenges agreed, a breakthrough in the field of prediction of protein folds. I evaluated the effectiveness of the DeepMind AlphaFold prediction tool by looking up at the Protein Data Bank Europe (PDBe) the entry for human 6 phosphogluconate dehydrogenase, which is the enzyme, but from sheep, whose X-ray crystal structure I worked on in my DPhil at Oxford University between 1974 and 1977. I compared the prediction (Helliwell 2021) of the human enzyme with the X-ray crystal structure of the sheep enzyme; the sheep enzyme entry was not available at the PDBe but sheep and human both being mammals meant that there are rather few amino acid sequence changes between them, and the protein fold should be very similar. These two folds, predicted and experimented, are compared in Figure 5.4.

To better understand this, I clarify: the terminologies of what is a protein fold, and what is a protein structure? The polypeptide chain fold is described by simply linking the alpha carbon in each peptide in the protein's amino acid sequence, one after the other. Then, where the individual atoms are is the structure, and as W L Bragg most famously said: *with crystallography, we can see atoms.*

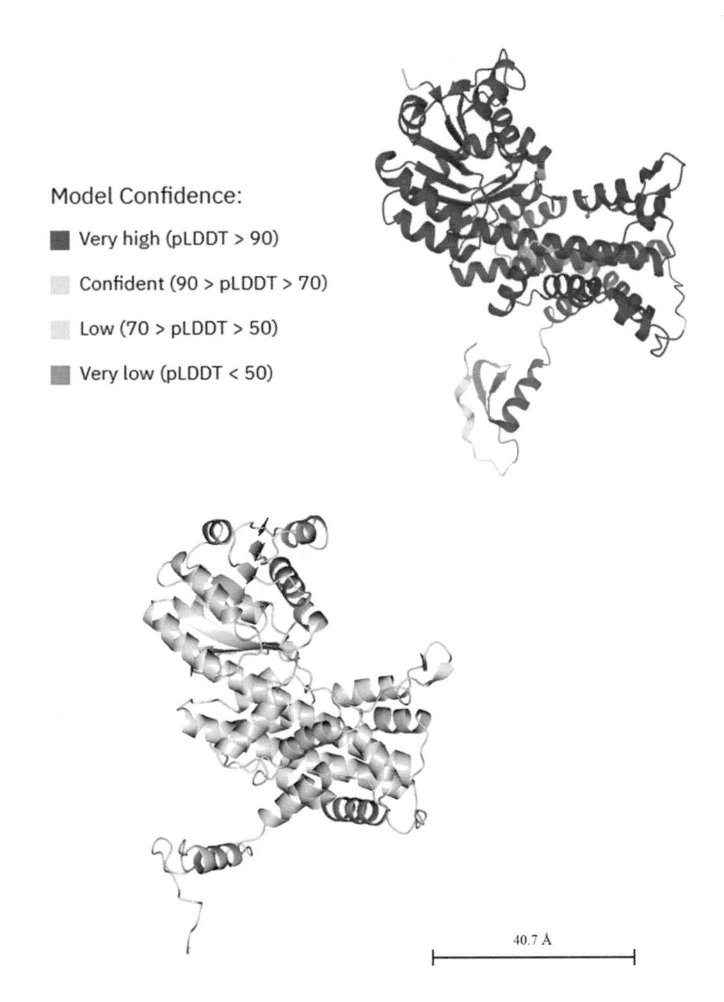

FIGURE 5.4 Top: The AlphaFold DB human 6 phosphogluconate dehydrogenase. This shows the ribbon diagram representing the polypeptide chain for the predicted 3D fold. Note the predicted "Model Confidence" at left, i.e., "Very high" throughout, apart from a short stretch of amino acids at bottom. This "Model Confidence" is described as *"AlphaFold produces a per-residue confidence score (pLDDT) between 0 and 100. Some regions below 50 pLDDT may be unstructured in isolation.*" Bottom: The X-ray crystal structure protein fold of sheep 6 phosphogluconate dehydrogenase (PDB code 2pgd) shown in a similar orientation for ease of comparison. This is the enzyme whose X-ray crystal structure determination was the focus of my DPhil in Oxford University between 1974 and 1977. The similarity of the folds predicted and from experiment are obviously striking. The top figure for details see Jumper et al. (2021) and which I viewed at the PDBe (Protein Data Bank Europe (PDBe) webtool) and took a web screenshot. The bottom figure I prepared using CCP4MG (McNicholas et al. 2011).

REFERENCES

Barber, B. (1987) Trust in science. *Minerva* 25, 123–134.

European Commission, Directorate-General for Research and Innovation. (2018) *Turning FAIR into Reality: Final Report and Action Plan from the European Commission Expert Group on FAIR Data*. Publications Office. https://data.europa.eu/doi/10.2777/1524.

Hackert, M. L., van Meervelt, L., Helliwell, J. R. & McMahon, B. (2016) *Open Data in a Big Data World: A Position Paper for Crystallography*. https://www.iucr.org/iucr/open-data.

Helliwell, J. R. (2019) Data science skills for referees: Biological X-ray crystallography. *Crystallogr. Rev.* 24, 263–272.

Helliwell, J. R. (2019) FACT and FAIR with Big Data allows objectivity in science: The view of crystallography. *Struct. Dyn.* 6, 054306.

Helliwell, J. R. (2021) Reaction to announcement of AlphaFold Database. *IUCr Newsletter* 29(2), 5.

Helliwell, J. R., Achari, A., Bloomer, A. C., Bourne, P. E., Carr, P., Clegg, G. A., Cooper, R., Elder, M., Greenhough, T. J., Shaanan, B., Smith, J. M. A., Stuart, D. I., Stura, E. A., Todd, R., Wilson, K. S., Wonacott, A. J. & Machin, P. A. (1981) Protein crystal oscillation film data processing: A comparative study. *Acta Cryst. A* 37, C311–C312.

Helliwell, J. R. & Massera, C. (2022) The four Rs and crystal structure analysis: Reliability, reproducibility, replicability and reusability. *J. Appl. Cryst.* 55, 1351–1358.

Helliwell, J. R., McMahon, B., Guss, M. & Kroon-Batenburg, L. M. J. (2017) The science is in the data. *IUCrJ* 4, 714–722.

Jumper, J., Evans, R., Pritzel, A. et al. (2021) Highly accurate protein structure prediction with AlphaFold. *Nature* 596, 583–589. https://doi.org/10.1038/s41586-021-03819-2.

Kroon-Batenburg, L. M. J. & Helliwell, J. R. (2014) Experiences with making diffraction image data available: What metadata do we need to archive? *Acta Cryst.* D70, 2502–2509.

Kroon-Batenburg, L. M. J., Helliwell, J. R., McMahon, B. & Terwilliger, T. C. (2017) Raw diffraction data preservation and reuse: Overview, update on practicalities and metadata requirements. *IUCrJ* 4, 87–99.

McNicholas, S., Potterton, E., Wilson, K. S. & Noble, M. E. M. (2011) Presenting your structures: The CCP4mg molecular-graphics software. *Acta Cryst. D* 67, 386–394.

Science International. (2015) *Open Data in a Big Data World*. International Council for Science (ICSU), International Social Science Council (ISSC), The World Academy of Sciences (TWAS), InterAcademy Partnership (IAP), Paris.

Strickland, P. R., McMahon, B. & Helliwell, J. R. (2008) Integrating research articles and supporting data in crystallography. *Learn. Publ.* 21(1), 63–72.

Tanley, S. W. M., Schreurs, A. M. M., Helliwell, J. R. & Kroon-Batenburg, L. M. J. (2013) Experience with exchange and archiving of raw data: Comparison of data from two diffractometers and four software packages on a series of lysozyme crystals. *J. Appl. Cryst.* 46, 108–119.

Taylor, R. & Kennard, O. (1986) Accuracy of crystal structure error estimates. *Acta Cryst. B* 42, 112–120.

van der Aalst, W. M. P., Bichler, M. & Heinzl, A. (2017) Responsible data science. *Bus Inf. Syst. Eng.* 59(5), 311–313.

Wilkinson, M. D. et al. (2016) Comment: The FAIR guiding principles for scientific data management and stewardship. *Sci. Data* 3, 160018.

CHAPTER 6

How Science Preserves Truth; The Editor As a Gatekeeper of Truth

There is one book (Fyfe et al 2022) that offers a 350-year (1665 to 2015) look back at the history of science publishing and which I have reviewed (Helliwell 2023). I opened my book review with the following comments:

> (This book) was inspired by the 350th anniversary of the Royal Society and the start of the first journal, Philosophical Transactions. At the celebrations, Sir Paul Nurse, Nobel Prize winner, observed that Henry Oldenburg had started the journal in 1665; Oldenburg had described four essential elements that its introduction would bring for scientists from all over the world, namely registration, verification, dissemination, and archiving. That said, the authors of the present book are clear that whilst the core facts are true, Oldenburg started it and the year was 1665, it is not the case that the needs of modern scientists and modern scientific publications were anticipated back then. More specifically, in a section entitled Going beyond the myths, it is pointed out that it was a further 87 years before Philosophical Transactions was to become an official publication. The authors conclude their opening chapter with the observation that their book matters not

DOI: 10.1201/9781003405399-6

> least because "the structures of academic publishing are being consciously renegotiated [today]".

In terms of scientific truth, the most significant section, and that I specifically commented on, was:

> Fig. 5.3 (Figure 6.1 here) shows the "advertisement" at the front of Volume 47 of the Transactions that the "truth of the facts, the soundness of the reasoning, or the accuracy of calculation" of any publication rested still "with the authors, not the Society." This was in contrast with the Paris Académie Royale's journal, where experiments were even redone to ensure the accuracy of a publication's results. This latter procedure could mean that publication was two to three years after submission. These details are of course very interesting in the context of modern debates on the reproducibility of science, for example the excellent report of the US National Academies of Science, Engineering, and Medicine (2019) or the UK Parliamentary Committee on Science and Technology enquiry (2022, report in preparation). One modern trend is a compromise between the Paris and London approaches of the 1750s, namely checking the underpinning data and software of a publication during prepublication peer review and thereby testing the soundness of its narrative [see also e.g Helliwell (2018) for a description of data science skills for referees of biological crystallography]. Post-publication peer review could of course include repeating experiments. In the 1750s an outcome was that London was accused of only publishing incremental science whereas Paris could publish leaps in discovery and even approve patents. That said, the Society claimed to publish papers of "importance." echos, then, of today where a journal's referees are routinely asked to evaluate the significance of a submission typically on a scale of 1 to 5. This can be a difficult thing to assess and is a source, upon rejection on such grounds, of authors feeling unfairly treated. These procedural issues, and matters of epistemological principle, impinged on the costs and benefits to the Society, which are described in detail with a bottom line that it would take two centuries before the Transactions showed a profit for the Society.

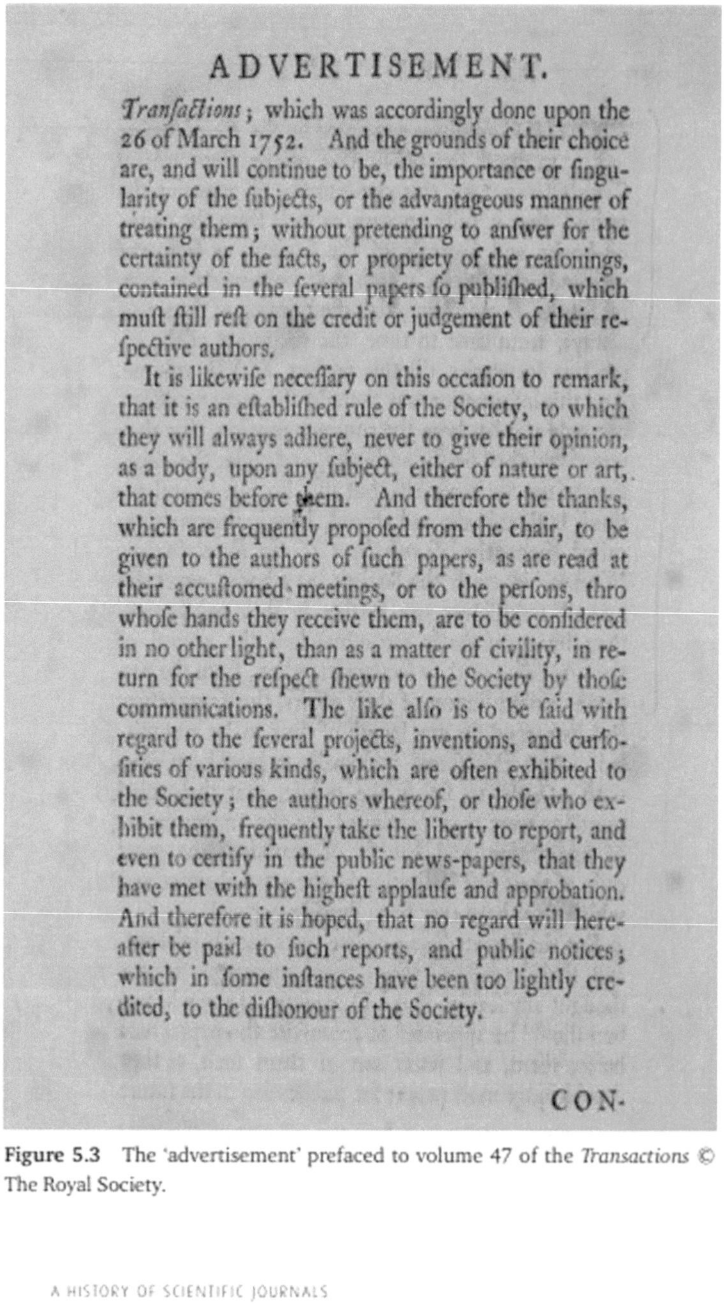

ADVERTISEMENT.

Tranſactions; which was accordingly done upon the 26 of March 1752. And the grounds of their choice are, and will continue to be, the importance or ſingularity of the ſubjects, or the advantageous manner of treating them; without pretending to anſwer for the certainty of the facts, or propriety of the reaſonings, contained in the ſeveral papers ſo publiſhed, which muſt ſtill reſt on the credit or judgement of their reſpective authors.

It is likewiſe neceſſary on this occaſion to remark, that it is an eſtabliſhed rule of the Society, to which they will always adhere, never to give their opinion, as a body, upon any ſubject, either of nature or art, that comes before them. And therefore the thanks, which are frequently propoſed from the chair, to be given to the authors of ſuch papers, as are read at their accuſtomed meetings, or to the perſons, thro whoſe hands they receive them, are to be conſidered in no other light, than as a matter of civility, in return for the reſpect ſhewn to the Society by thoſe communications. The like alſo is to be ſaid with regard to the ſeveral projects, inventions, and curioſities of various kinds, which are often exhibited to the Society; the authors whereof, or thoſe who exhibit them, frequently take the liberty to report, and even to certify in the public news-papers, that they have met with the higheſt applauſe and approbation. And therefore it is hoped, that no regard will hereafter be paid to ſuch reports, and public notices; which in ſome inſtances have been too lightly credited, to the diſhonour of the Society.

CON-

Figure 5.3 The 'advertisement' prefaced to volume 47 of the *Transactions* © The Royal Society.

168 A HISTORY OF SCIENTIFIC JOURNALS

FIGURE 6.1 The "advertisement" prefaced to volume 47 of the Transactions of The Royal Society. From Fyfe et al. (2022), copyright from The Royal Society.

I commented on Chapter 13 as follows:

> Chapter 13 elaborated further on these important points of reproducibility and the need for the underpinning data. The chapter is entitled Why do we publish? 1932–1950. This question was posed by Sir William Henry Bragg in his presidential address of 1938. The question impinged on the Society's expectations of a submitted paper. To be publishable the paper should "contain methods or results of critical importance" and be "of value to others than specialists in the particular subject." These criteria would "avoid [the] unnecessary expense" of publishing routine work. Where to draw the boundary was obviously difficult, as illustrated by W. H. Bragg trialling the archiving of data for X-ray crystallography results where "even Bragg admitted that most of these data would only be needed by those 'very few readers' who wanted to 'check the detail of the work.'" So having the underpinning data was a good practice but the general reader probably would not have the expertise to redo the calculations of the authors. Obviously, specialist journals would have a major advantage in ensuring reproducibility of the studies that they published.

In the field of crystallography, as an example of a science domain with specialist journals serving it, a leader has been the journals of the International Union of Crystallography (IUCr) itself. In 1948, the IUCr was formed and launched its own learned society journal, Acta Crystallographica. I am proud to say that I was its Editor in Chief for three triennia, the maximum allowed, from 1996 to 2005. Coming from biological crystallography, I immediately noticed how much more advanced the chemical crystallography domain was in its peer review of the article with data. The Cambridge Structure Database had been launched in 1965. Thus, over the decades, the chemical crystallographers developed a deep, community-wide knowledge of how to validate their chemical crystal structures. These checks became enshrined in a Crystallographic Information Framework. This is maintained by the IUCr's Committee for the Maintenance of the CIF Standard and reports to the Executive Committee of the IUCr. Also, in this community-wide sharing of their refereeing of data-with-article experiences, the community itself has set standards and helped expand the checks, of which there are now over 400 in number; this is known as checkcif (http://checkcif.iucr.org/).

Educational training events are also important. Two of the most recent are shown in these images from the IUCr website (Figure 6.2 Top and 6.2 Bottom):

So, education has a vital role to play in the scientific process so that it can be trusted. There are efforts by disparate science communities to introduce new terms to ensure trust in science. These new terms have merit for discussion in crystallographic teaching commissions and possible adoption by crystallographers too. We have published a recent teaching and

WORKSHOP ON *DATA SCIENCE SKILLS IN PUBLISHING: FOR AUTHORS, EDITORS AND REFEREES*

Organized by

IUCr Committee on Data

SUNDAY AUGUST 18 2019

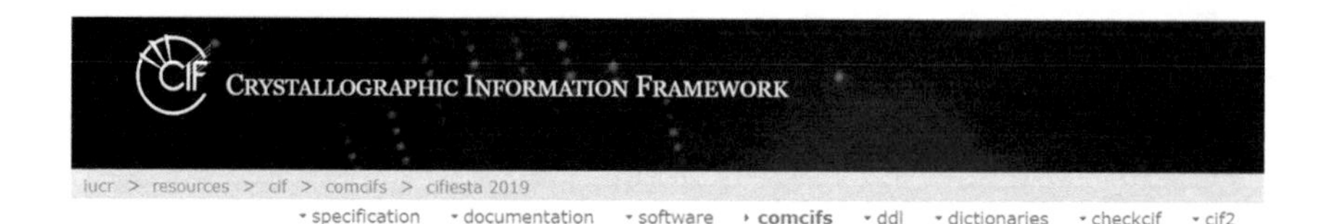

CRYSTALLOGRAPHIC INFORMATION FIESTA

CIFiesta

In collaboration with the IUCr Committee on the Maintenance of the CIF Standard (COMCIFS) and Committee on Data (CommDat), the Associazione Italiana Cristallografia held in autumn 2019 the first CIFiesta, an intensive course in Crystallographic Information, covering the extraction, dissemination and use of scientific knowledge from the structure determination experiment to database-driven discovery. Photographs of the event are in the IUCr **photo gallery** and course materials and schedule may be found on this page. An account of the school appears in the ***IUCr Newsletter***, Vol. **27**, No.4.

Motto of the School: > ***"Know your data - Trust your data - Share your data"***

FIGURE 6.2 Top and bottom. Recent educational training events in crystallography involving information and data from www.iucr.org.

education article on this topic where we envisage a students' discussion seminar on "trust in science and the role of crystallography" where the students could explore firstly the domain of crystallography and then, more broadly, history of science examples (Helliwell and Massera 2022). Loss of trust reduces the confidence inside a community and as well can disparage the way that a community is perceived from outside. Trust in science is built up from different facets. To that end, the crystallographic community has for many decades used the word "reliability" as exemplified by its R-factors as well as other metrics. Other science communities and policy bodies have new terminologies such as FAIR and FACT, as described in Chapter 5. These developments connect with efforts to improve reproducibility and replicability in science, in general as exemplified by the US National Academies of Sciences, Engineering, and Medicine, which published in 2019 on the Reproducibility and Replicability of Science (US National Academies of Sciences, Engineering and Medicine 2019). Independently, the crystallography community has developed and indeed led the way in introducing a distinct crystallographic information framework (CIF) of clear ontologies within a CIF file (Hall and McMahon 2016). The International Union of Crystallography has a Committee for the Maintenance of the CIF Standard (https://www.iucr.org/resources/cif/comcifs), established in 1993. Central to this approach is a check of the CIF file; checkCIF reports on the consistency and integrity of crystal structure determinations reported in CIF format. Similarly, any Protein Data Bank (PDB) deposition involves an extensive advisory PDB validation report (https://www.wwpdb.org/validation/validation-reports) assessing numerous indicators of correctness against the processed diffraction data and expected molecular geometry values. In conclusion, crystallography is a discipline where community-agreed processed diffraction data and model validation checks are routinely made. Although this system is not perfect, it provides the best chance for ensuring reliability and thereby, trust in what we do. Most importantly these standards preserve our community. We think however that the terminologies from other areas of the sciences could usefully assist the crystallographic community in its policies, such as in journal notes for authors, as well as how we engage with the public and students (Helliwell and Massera 2022).

Through the leadership of the chemical crystallography community's style of refereeing of article with underpinning data, comprising its general checks (http://checkcif.iucr.org/) and the specific structural chemistry expertise of an editor's chosen referees, the IUCr President in 2022 announced in the IUCr Newsletter the IUCr Journals' Management Board decision to adopt this procedure in all crystallography's domains. See https://journals.iucr.org/services/datasharingpolicy.html and to quote this in full:

> The IUCr supports the principles of transparency and openness in scientific research. This means making research data available to facilitate the reproduction, validation, reuse and reinterpretation of research findings, leading ultimately to more effective research discovery, i.e. Findable Accessible Interoperable Re-usable (FAIR).
>
> **IUCr journals mandate data sharing and peer review of crystal structure data**
>
> The IUCr has adopted a data-sharing policy that requires the crystal structure data supporting the results in an article to be peer reviewed and archived either with the IUCr or in an appropriate public repository. Full details of the requirements are given in the Notes for authors. For articles describing scripts or program code, whenever possible these and any other documentation necessary for reproduction of the published results should also be publicly archived. This policy applies to all IUCr journals. (In rare cases exceptions may be granted by the editors, for example when the sharing of data may compromise ethical standards or legal requirements.)
>
> Published articles contain links to the data either on the IUCr site or in the repository together with links to other supporting information. All relevant accession, reference or identification codes, or other persistent identifiers such as DOIs are included in articles.
>
> It is the practice of IUCr journals to provide free access to all supplementary materials and supporting data files deposited with a published article. When the data are reused, we ask that proper attribution is given to the associated source article.
>
> **DATA REPOSITORIES**
>
> The IUCr recommends that authors deposit data sets in standard, subject-specific public repositories. A list of repositories relevant to crystallographic data is available at https://www.iucr.org

/resources/data/databases. Additional repositories may be found by visiting https://www.re3data.org or https://fairsharing.org, which list registered and certified data repositories.

DATA REFERENCES

Identification of individual structures in an article by use of database codes should be accompanied by a full citation of the original literature in the reference list.

Where authors have used a publicly available data set from a repository this should be cited in the published article according to the following style:

Authors (Year). Data set title. Persistent identifier.

An example citation is:

Brink, A. & Helliwell, J. R. (2019). Raw diffraction images. Formation of a highly dense tetra rhenium cluster in a protein crystal and its implications in medical imaging. https://doi.org/10.5281/zenodo.2874342

REFERENCES

Brink, A. & Helliwell, J. R. (2019) Formation of a highly dense tetra rhenium cluster in a protein crystal and its implications in medical imaging. *IUCrJ* 6, 695–702.

Fyfe, A., Moxham, N., McDougall-Waters, J. & Røstvik C. M. (2022) *A History of Scientific Journals: Publishing at the Royal Society, 1665–2015*. University College London Press. https://doi.org/10.2307/j.ctv2gz3zp1.

Hall, S. R. & McMahon, B. (2016) The implementation and evolution of STAR/CIF ontologies: Interoperability and preservation of structured data. *Data Sci. J.* 15, 3.

Helliwell, J. R. (2018) Data science skills for referees: I biological X-ray crystallography. *Crystallography Reviews*, 24:4, 263–272, DOI: 10.1080/0889311X.2018.1510878

Helliwell, J. R. & Massera, C. (2022) The four Rs and crystal structure analysis: Reliability, reproducibility, replicability and reusability. *J. Appl. Cryst.* 55, 1351–1358.

National Academies of Sciences, Engineering and Medicine. (2019) *Reproducibility and Replicability in Science*. The National Academies Press, Washington, DC.

CHAPTER 7

Post-Publication Peer Review

In my experiences as Editor in Chief of the International Union of Crystallography (IUCr) Journals from 1996 to 2005, I gained an overview of all the subject domains of crystallography. As I remarked in the previous section, I was very impressed by the chemical crystallography community's efforts to ensure their publications and linked data were truly a version of record. My personal research expertise was in protein crystallography, within which I had undertaken research and development in synchrotron radiation instrumentation, methods, and applications. I was fortunate to have had the chance to develop protein crystallography beamlines at the world's first dedicated synchrotron radiation source, the SRS. Then, subsequently, at the University of Manchester, I developed some specific structural chemistry and biology research themes of my own. Although I set myself a wide research remit, it was limited, nevertheless. My experiences as Editor in Chief of IUCr journals from 1996 to 2005 greatly broadened my perspectives.

Listening to feedback from the crystallographic community was very important to me in this role as Editor in Chief. Every three years we have a world congress and general assembly of crystallography, and within these congresses we could hold open commission meetings. My 2005 Congress abstract, coauthored with the IUCr journals Managing Editor Peter Strickland, is reproduced in Figure 7.1 below as a simple screenshot. Each section editor of the IUCr journals had to do likewise, present a summary,

DOI: 10.1201/9781003405399-7

OPEN COMMISSION MEETINGS

OCM01 Commission on Journals (I)
Coordinator: John R. Helliwell

OCM01.24.1
Acta Cryst. (2005). **A61**, C125
Overview of IUCr Journals
John R. Helliwell[a,b], Peter R. Strickland[c], [a]*School of Chemistry, The University of Manchester, M13 9PL, UK.* [b]*CCLRC, Daresbury Laboratory, Warrington WA4 4AD, UK.* [c]*IUCr Journals, 5 Abbey Square, Chester CH1 2HU, UK.* E-mail: john.helliwell@manchester.ac.uk

Comparing the year ends of the last three triennia, 11728 journal pages were published in 2004, compared with 9215 in 2001 and 7937 in 1998. This increase in the number of pages has been accompanied by a major reduction in publication times for all sections of *Acta Cryst.*, *J. Appl. Cryst.* and *J. Synchrotron Rad.* In the most recent triennium, an electronic submission system was introduced and has been very popular with Co-editors and authors. The journals continued to be the most cited in crystallography; *Acta Cryst. B* currently has the highest impact factor (3.643). The overall withdrawal plus rejection rate for *Acta Cryst.* was 26% in 2004, up compared with 2002 (18%). For *Acta Cryst. A*, special issues based on workshop 'hot topics' have been introduced, similar to the strategy in recent years for *J. Synchrotron Rad.* An open-access option was introduced for authors in 2004; grants have allowed all UK papers to be published open access in 2004/2005. *Acta Cryst. E* has been very successful in attracting ever increasing numbers of electronic structure reports. *Acta Cryst. C* is increasingly the home of the most important and high-quality crystal structure communications; its impact factor rising from 0.571 in 2001 to 0.828 in 2003. In 2004 we launched *Acta Cryst. F: Structural Biology and Crystallization Communications* (Editors H. Einspahr and M. Guss); extensive work with the PDB has been made by H. Einspahr on the streamlining of deposition-to-publication methodologies. Finally, a review is currently being made of Education papers within IUCr Journals.
Keywords: journals publishing, commission on journals, overview

FIGURE 7.1 An important form of post-publication peer review is the open meeting where the community of a journal can question the editor of a journal, as illustrated here with my own abstract at the World Congress of Crystallography held in Florence in August 2005. Reproduced with the permission of IUCr journals.

and listen to feedback from the community of authors and readers that they served.

In these open commission meetings, it is fair to say that there were not major criticisms made of the IUCr journals. However, in this period of 1996 –2005 scientific publishing had not really started the process of reinventing itself, as referred to in Chapter 6 . In my final remarks as Editor in Chief to the IUCr General Assembly of 2005 in Florence though I stated that the open science movement's emphasis on journals becoming "open" i.e, free to readers, would have significant consequences for authors who could not afford to pay the article processing charges that would replace the journal subscriber financial model. In its 2023 open science summary document, UNESCO refers to "Article processing charges being one of the most prominent unintended consequences of open science." Were the architects of article processing charges (APCs) really so naive that they did not realize that barriers to authors of APCs would be a big problem for many, and not only in the global south, e.g., retired scientists everywhere. There is now a campaign for a "free to readers and free to authors publishing model"; this is referred to by UNESCO as diamond open access. It is also sometimes referred to as platinum open access. This is not as impossible as it initially seems. Journals like eLife are financed by the Wellcome Trust and thereby make it viable for "free to readers and to authors." The journal eLife's scope is restricted to medical or biomedical subject areas. In terms of post-publication peer review, the focus of this part of my book, clearly no one is excluded from reading a publication or authoring a publication on a given research study in eLife. For good measure, eLife also publishes the accompanying referees' reports and the responses of the authors to these, as well as the editor's decision letters. When there is a publication in eLife that I am interested in, I do find all these reports, and the final article, very interesting. I should also say that whilst I note the enthusiasm of funding agencies for preprint servers, the filtering step by a journal, ie publishing solely those articles which are passed by pre publication peer review of that journal, to a manageable number of articles that I can read, I greatly appreciate. So, the recent decision by eLife to not reject any article is unfortunately removing that quality filtering step that I appreciate in carefully managing my limited time.

To come back to the more usual style of journals, how are critique articles handled, and how many are there? In crystallography, a significant theme of the 1990s was the systematic vetting of chemical crystal structures where the researchers had chosen a too low a symmetry space group. There are 230 three-dimensional space groups that are possible mathematically, and so it can be a difficult decision step for a researcher to make in any new crystal structure to definitively assign a crystal's space group. A too low a symmetry would offer a sort of comfortable choice but can lead to instability in a refinement of the atomic model against the diffraction data. It is always better to choose a space group where all symmetry elements in the crystal layout have been identified. A specialist in this topic of spotting space group errors was Professor R E Marsh. His efforts even led to a new verb, to be "Marshed." In my term as Editor in Chief, one such publication involved 60 crystal structures needing their space groups correcting was Marsh et al. 2002. Whilst somewhat technical, I quote their abstract in full:

> Some 60 examples of crystal structures are presented which can be better described in space groups of higher symmetry than used in the original publications. These are divided into three categories: (A) incorrect Laue group (33 examples), (B) omission of a center of symmetry (22 examples), (C) omission of a center of symmetry coupled with a failure to recognize systematic absences (nine examples). Category A errors do not lead to significant errors in molecular geometry, but these do accompany the two other types of error. There are 19 of the current set of examples which have publication dates of 1996 or later. Critical scrutiny on the part of authors, editors and referees is needed to eliminate such errors in order not to impair the role of crystal structure analysis as the chemical court of last resort.

The important role of crystallographic databases in helping police such cases is noted by Marsh et al. 2002 in their acknowledgements:

> The staff of the CCDC (Cambridge Crystallographic Data Centre) have been very helpful in providing information.

A situation in 2007 arose way beyond simple critique, involving the discovery of deliberately falsified chemical crystal structures. These cases involved the utilization of bona fide intensity data of correctly determined crystal structures reported in the literature to create new, fantasy structures. There were about 70 of these. Compared with the total number of chemical crystal structures in the Cambridge Structure Database, this was a tiny number.

The detection of these was done by Ton Spek of Utrecht University and his software PLATON [Spek 1990, 2003, 2020]. His role was much appreciated by the chemical crystallography community; the Main Editors of Acta Cryst E in their 2010 editorial [Harrison et al. (2010)] remarked:

> These problems were first discovered by Ton Spek during testing of the checking programs for the journal. Testing is routinely carried out using cifs and structure-factor files from back issues of *Acta Crystallographica Sections E* or *C*. Initially, unexplained Hirshfeld rigid-bond alerts and unusual metal–ligand donor-atom distances led to the discovery that metal atoms had been transposed and that more than one structure had been "determined" using identical sets of data. Investigation of these cases sparked a search of papers written by the correspondence authors involved.

Two main strategies were used by the falsifiers:

(i) metal exchange in coordination complexes bearing the same ligand (*i.e.*, the structure of a zinc complex would be used to obtain similar complexes with copper, cobalt, nickel *etc.*) and compounds (especially with lanthanides);

(ii) element exchange in organic compounds (for instance, CH_2 groups were replaced by NH_2 or O and vice versa; OH groups were replaced with F atoms, and so on).

The editors went on to state (Harrison et al. 2010):

> When we discussed the events with the Editors of other journals in the Acta family, they expressed amazement, because, like us, they assumed that it was almost inconceivable that a fake crystal structure would be submitted for publication. Sadly, that has proven not to be the case and we must now take stock and decide what steps are needed to prevent further scientific fraud. To that end, the checkCIF validation software is being improved continuously and provides an exhaustive assessment of data and structural quality and consistency. It is also noteworthy to point out that the current problems could not have been easily discovered without the availability of the structure-factor files; it will become increasingly important for all journals reporting crystal structures to make sure that they require authors to supply such data in future.

> Finally, nothing can replace the sceptical (but fair) assessment of an experienced Co-editor. While it is impossible to give absolute guarantees that such a situation will not happen again, we feel that the journal, its Editors, Co-editors and the Chester staff are now far better prepared to identify and challenge any further attempts to publish anything other than articles reporting genuine structural investigations in our journal. It is a strength of crystallography that fraudulent practices can be identified, even retrospectively, by diligent archiving of data and checking such as that carried out for the Union's journals. We thank Ton Spek, George Ferguson and the IUCr Editorial Staff for all their input and assistance.

In more recent times, there has been the papermill issue (Chawla 2022). That one was related to the "invented text" of papers claiming medical and pharmaceutical novelties. Ton Spek is again the go-to person to get the authoritative view on the issue. To me, he replied, "*The crystallography was not that bad in most cases. I looked at a number of those papers. The IUCr journals' articles were not within the set of 1000 or so faked papers.*"

Clearly, post-publication peer review is a necessity, even in a domain like chemical crystallography, which I still regard as an exemplary effort in its pre-publication peer review procedures to make a publication the true and definitive version of record.

REFERENCES

Chawla, D. S. (2022) 800 Crystallography-Related Papers Appear to Stem from One Paper Mill. *Chemistry World*. https://www.chemistryworld.com/news/800-crystallography-related-papers-appear-to-stem-from-one-paper-mill/4015589.article.

Harrison, W. T. A., Simpson, J. & Weil, M. (2010) Editorial. *Acta Cryst. E* 66, e1–e2.

Marsh, R. E., Kapon, M., Hu, S. & Herbstein, F. H. (2002) Some 60 new space-group corrections. *Acta Cryst. B* 58, 62–77.

Spek, A. L. (1990) PLATON, an integrated tool for the analysis of the results of a single crystal structure determination. *Acta Cryst. A* 46, c34.

Spek, A. L. (2003) Single-crystal structure validation with the program PLATON. *J. Appl. Cryst.* 36, 7–13.

Spek, A. L. (2020) CheckCIF validation ALERTS: What they mean and how to respond. *Acta Cryst. E* 76, 1–11.

CHAPTER 8

The Issue and Challenge of Archiving All Data

THE VIEWPOINTS OF TWO key players in the whole research enterprise, the funders and the publishers, are not identical. Figure 8.1 from an F1000 webinar neatly summarizes this. The publisher is not driving the need for "all data" to be preserved. The concern of the publisher is that a person interested in one of their publications can reproduce the statements made in the publication. The implicit assumption by the publisher is also that the study is "typical." The funder, on the other hand, may have concerns about what the team leader has selected for publication. This concern may have more of an edge to it if it is a study aiming at a medical application, say, than a rather erudite topic of investigation. There is a third stakeholder in all this. In quite a few fields of science, the experimental equipment has become so complex that an individual team leader cannot hope to host that apparatus in their own laboratory. Instead, "central facilities" are established. These include synchrotron radiation, X-ray laser, and neutron facilities. Also, in astronomy, telescopes are centrally run, as are particle and nuclear physics research facilities. These facilities may be especially concerned if not all the measured data lead to publications, as that would represent inefficiency in their operation if that were the case.

These issues are at a special focus today, as in each region of the globe, funders are keen to see "open science clouds." Thus, as an example, the European Union is establishing a European Open Science Cloud, likewise China and the USA. Africa has the Open Science Platform. It is not

DOI: 10.1201/9781003405399-8

clear how the governance of these open science clouds will operate. To ensure a uniformity of approach and best practice is adopted in each, the International Science Council's Committee on Data, "CODATA," has established a Global Open Science Cloud (GOSC) as a federation approach to these developments. The coordination office for this GOSC is based in Beijing. The first international symposium on open science clouds was hosted in Beijing in September 2023 (https://codata.org/international-symposium-on-open-science-cloud-2023/). I presented a talk at this symposium on our progress with case study 5 https://codata.org/initiatives/decadal-programme2/global-open-science-cloud/case-studies/diffraction-data/.

I will conclude this Chapter 8 by drawing the analogy of "all data" with the "whole truth." In a legal trial, the witness swears on oath to tell "the truth, the whole truth, and nothing but the truth." But lawyers seem to use certain lines of questioning to develop an argument. Witnesses are encouraged by the lawyer to simply answer the questions asked and not say much more. This is enforced by the presiding judge, at least in courtroom TV dramas that we watch.

Now back to science. A quite famous case in 2023 was the article published in Nature (Brown et al. 2023) on climate change as a cause of the increase in forest fires. Dr. Magdalena Skipper, Editor in Chief of Nature, wrote to The Times (Saturday, 16 September 2023) concerning an article

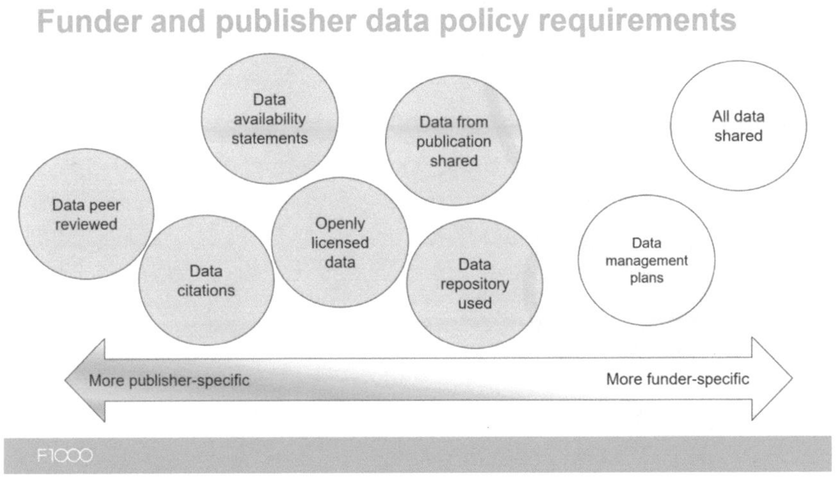

FIGURE 8.1 A comparison of funder and publisher data policy requirements. Reproduced with the permission of F1000, Taylor and Francis.

from the previous week (Saturday, 9 September 2023) by one of Nature's authors Patrick T Brown. He had argued that "*to get published in a prestigious journal, climate researchers might find they have to omit facts and ignore context.*" Dr. Skipper argued that the article of Dr. Brown and his coauthors' "*focus was questioned by fellow climate scientists who reviewed the paper before publication and Dr Brown himself argued (persuasively) not to include factors other than climate change. Plainly, the choice of focus was not Nature's but the author's.*" Dr. Brown had stated plainly the previous week in The Times that "*I am a climate scientist. And while climate change is an important factor affecting wildfires over many parts of the world, it isn't close to the only factor that deserves our sole focus.*" This is an important argument between author and editor and on an important issue. A key point is missing in this dialogue, which is that the editor is the gatekeeper of truth and in this case, a narrow version of the truth, what we can call "not the whole truth," is to my mind clearly unacceptable. Usually, the critique of a published article comes from scientists other than the original authors!

REFERENCE

Brown, P. T., Hanley, H., Mahesh, A., Reed, C., Strenfel, S. J., Davis, S. J., Kochanski, A. K. & Clements, C. B. (2023) Climate warming increases extreme daily wildfire growth risk in California. *Nature*. https://doi.org/10.1038/s41586-023-06444-3.

CHAPTER 9

Conclusions and Future Outlook

TRANSPARENCY OF A RESULT requires the underpinning data to be made open along with the publication and is a key component of open science. The United Nations and UNESCO have made this a major focus in the past two years. The USA National Academies of Science, Engineering, and Medicine (NASEM) published their excellent 2019 report on reproducibility and replicability in science, within which open data underpinning publications is key. Furthermore, though, the USA NASEM along with the Nobel Foundation are very concerned about misinformation and disinformation in science as a threat to scientific truth.

The UNESCO Recommendation on Open Science was adopted by the General Conference of UNESCO at its 41st session on 23 November 2021. The recommendation affirmed the importance of open science as a vital tool to improve the quality and accessibility of both scientific outputs and scientific processes, to bridge the science, technology, and innovation gaps between and within countries and to fulfill the human right of access to science. Details are here: https://www.unesco.org/en/open-science?hub=686

UNESCO convened five ad-hoc working groups focusing on key impact areas, bringing together experts and open science entities, organizations, and institutions, according to their field of activity and expertise on:

(i) Open Science Capacity Building

DOI: 10.1201/9781003405399-9

(ii) Open Science Policies and Policy Instruments

(iii) Open Science Funding and Incentives

(iv) Open Science Infrastructures

(v) Open Science Monitoring Framework

These have each had three virtual events. Recordings are available here: https://www.unesco.org/en/open-science/implementation

Having attended these working groups, I prepared reports for the Editor in Chief and Managing Editor of International Union of Crystallography (IUCr) Journals and the IUCr CEO. A key output of the working groups assembled by UNESCO is a helpful toolkit https://www.unesco.org/en/open-science/toolkit. Within this toolkit is a checklist for publishers: https://unesdoc.unesco.org/ark:/48223/pf0000383327.

A linked event to this UNESCO initiative and its various events was a three day hybrid event on open science organized by the UN Library (https://www.un.org/en/library/OS23) with a focus on accelerating the UN's sustainable development goals and democratizing the record of science.

On behalf of IUCr, specific inputs were provided into these events on (i) the importance of global collaborative sharing of instrumentation [citing our own contribution Warren et al 2008] and (ii) a fine example of open science in practice being the European Synchrotron Radiation Facility (ESRF) heritage database for palaeontology, evolutionary biology, and archaeology (http://paleo.esrf.eu/). Basically, so much of these data can be measured quickly that Europe's palaeontologists decided to share all their data with the whole world's palaeontologists to analyze the data as promptly as possible.

A specific conclusion from the UN/UNESCO deliberations features a growing importance of, and seeking funds for, the prioritising of diamond open access for publications. Diamond open access refers to a scholarly publication model in which journals and platforms do not charge fees to either authors or readers e.g. see 202203-diamond-oa-action-plan.pdf (https://www.scienceeurope.org/media/t3jgyo3u/202203-diamond-oa-action-plan.pdf).

Open science has wider definitions than this, however, such as making data open after an embargo period of typically three years, even if the data in a taxpayer-funded project has not, by the time of the three years elapsed since measurement, reached a publication. This is a central policy pillar of the European Photon and Neutron Open Science Cloud (see e.g. the data policy of the European Synchrotron Radiation Facility https://www.esrf.fr/files/live/sites/www/files/about/organisation/ESRF%20data%20policy-web.pdf). In the USA, by contrast, even though its National Academies of Sciences, Engineering, and Medicine (2018) report seeks to achieve for the USA "Open Science by Design: Realizing a Vision for 21st Century Research," it places an important boundary on the degree of openness, from which I quote:

> **Sharing prior to the point of publication is up to the researcher, who is in full control of the decision of when to share.**

The sentence above that I have placed in bold is for emphasis because, just who does own research data seems to involve a different policy in the USA versus that in Europe, which is very interesting. See Helliwell (2022) https://forums.iucr.org/viewtopic.php?f=39&t=445 *"Just who does own research data?"*

Also, open science has its critics in some aspects, such as Leonelli (2023) based on a lack of inclusiveness:

> "The pursuit of truth requires discrimination, and so does the practice of openness. Researchers are constantly making hard choices. Among those choices are decisions around what to make open, to whom, when and for which reasons."

To understand the enthusiasm for openness in science is straightforward, in my view, for published work. The open science movement, as Leonelli (2023) and others call it, is aware of the resource-imposed limits on inclusiveness by seeking to adopt a diamond open access publication model of free for authors as well as the more common free for readers, Gold or Green Open Access (United Nations Open Science Conference 2023 https://www.un.org/en/library/OS23). For unpublished data, there is a developing concern about large quantities being unpublished, making those "dark data" (*i.e.*, currently never seeing the light of day data). If those data are made open as well then they may become more useful.

Secondly, the question may reasonably be asked: why have the researchers selected just this portion of measured data for their publication? Making all the data available may well increase the trust of others in a published result, and that it is "typical," not somehow highly selected, and possibly "atypical."

Indeed, what can be done with data if all of it is released to others beyond the original measuring team? Furthermore, is this a domain of data where machine learning and artificial intelligence would finally allow benefit to be gained from the investments made by funding agencies and the measuring teams in measuring those dark data? These are unanswered questions at present, as the necessary policies are only recently coming into force.

Leonelli (2023) also introduces an emphasis on connectedness, as well as inclusivity, that should define open science. Leonelli illustrates the general problem that a database's entries are lodged by experts, validated, and ensured with complete metadata by the database's staff, but are used much more widely than those methodologically qualified experts. The efforts of the databases to ensure that their entries are FAIR and FACT for all are then to be admired.

At the bottom line, openness of the underpinning data for a publication is essential for reproducibility of a study. Furthermore, trust is increased in a published study if it is made clear whether the selected data are typical or atypical.

Since the first ever X-ray crystal structure (Bragg 1913) crystallographers have strived to make the linking of a publication to its underpinning data happen. This started systematically in 1965 with the foundation of the Crystal Structure Database by the Cambridge Crystallographic Data Centre, now comprising >1.2 million crystal structures. Then in 1971, the Protein Data Bank was launched, now with >200,000 protein crystal structures. The International Center for Diffraction Data (the ICDD) has around half a million powder diffraction data patterns as entries. There are several other important databases. It is a big topic, and a review of the landscape of all the crystal structure and diffraction databases is that of Bruno et al. (2017). The other domain of the release of data that has not led to publication is complicated, as I have described, but may lead to major benefits of an improved use of limited funding resources. At any rate, all stakeholders in a research study can surely agree on a data management **and** sharing plan before anything is measured.

At risk of finishing this book on a sombre rather than optimistic note, it is clearly of great interest to all of us that the USA NASEM, along with the Nobel Foundation, are very concerned about misinformation and disinformation in science as a threat to scientific truth, and to society at large.

Specifically, if trust in truth breaks down, society and democracy itself are at grave risk. As I remarked in my Preface, my book is not a treatise about falsification of results, although I have described some specific instances from chemical crystallography. There is also misinformation or disinformation which is an interesting middle ground. So, let's explore misinformation or disinformation, as it was important enough for the Nobel Foundation and the USA National Academies of Science, Engineering, and Medicine (2023) held a three-day event in May 2023 entitled "Truth, Trust and Hope." The recordings are here:

https://www.youtube.com/playlist?list=PLJE9rmV1-0uD_jpBJp4oJLnitdZoyAcpI

On day 1, there was a wide range of examples of misinformation and disinformation, with a focus on the latter. The much-referred-to topics, as examples, were disinformation on climate change and COVID-19 vaccination, as well as the more general issues of manipulation of opinion against journalists and anti-government "activists". On day 2, there were breakout sessions exploring solutions to these issues. The report back from panels representing these breakout groups was also very interesting. One solution, which I found promising, was to deal with cases involving a "diversity of views of scientists" situations. Here, an expert panel can be consulted, and a variance or confidence can be established to help the public evaluate an issue. A second solution to the mis/disinformation crisis would be establishing an authoritative check of journals, run, e.g., by the USA NASEM, so that predatory journals could be exposed for the public and politicians to know if published work in those journals was likely to be truthful or not. Everyone could then be reticent to quote such published work. Thirdly, a solution could be to use AI to monitor which media or individuals in the (social) media are spreading mis/disinformation. Then these sources could be downweighted. A fourth solution would be validating the statements of leaders of communities. This reminds me of, in the UK, the BBC's fact- checking of speeches by political leaders. Another breakout group emphasized their support of the 10-point plan (https://peoplevsbig.tech/10-point-plan) to preserve the role of journalists in exposing falsehoods and preserving truth. Prof. Asa Wikfors of Stockholm University made the important point that "*Legislation by Government against mis/disinformation is a tricky issue, due to freedom of speech, the cornerstone of democracies*" and went on to say that… "*The answer is citizens' assemblies, i.e. deliberative polling.*" This sounds like the Swiss polls-on-issues method, and which are held monthly. James Fishkin of Stanford University: added that *"Deliberative assemblies are not like juries, which must come to a verdict."*

He is the author (Fishkin 2018) of https://www.amazon.com/Democracy-When-People-Are-Thinking/dp/0198820291.

At the start of day 3, Marcia McNutt, President of the USA National Academies of Science, Engineering, and Medicine, stated that "*The past two days stressed to me that education is so important and in particular to train young people in critical thinking skills.*" and Vidar Helgersen of the Nobel Foundation stated that "*The past two days we have learnt that lack of truth is destroying our democracies. Today has nine sessions for solutions to this challenge. The teacher is so important for the future of our societies.*" A very interesting final question was put to the Future of Education Panel: "*Why is teaching critical thinking in schools better than teaching facts?*" Several answers were put forward by the Panellists to this question:

> 1st answer: It must be both. But it would be better to teach the STEM examples which have transformed our view of the world.
>
> 2nd answer: At Planet Word we are teaching, before critical thinking, a love of literacy.
>
> https://planetwordmuseum.org/about/#:~:text=Planet%20Word%20will%20inspire%20and%20renew%20a%20love,a%20solid%20understanding%20of%20language%20arts%20and%20science.
>
> There were two answers mentioning ChatGPT expressing concern that it is not able to discover new truths.
>
> A final answer was that: "*Relevance is a big factor in whether we seek to learn something new.*"

The second final question put was, "*Why is this a transformational period for education?*"

Two panellists answered that: "*Teaching science for civic engagement is much more important these days than teaching science for workforce preparation.*"

A 3rd panellist answered that: "*There have always been big challenges, such as living under the threat of nuclear wars.*"

The USA NASEM with Nobel Foundation Symposium on Truth, Trust, and Hope is a milestone. The disinformation levels, "fake science," have reached alarming levels. Aspects of this are not new though. In 1954, the book by Darrell Huff (Huff 1954) was published. It sold well, being reprinted many times and a second edition was published in 1973 (see review by Herne 1973). As Herne (1973) remarked this book is aimed at:

> men in the street who don't know immediately how to view graphs with suppressed zeros or with distorting scales, what precision is nor when differences between numbers are meaningful, who know even less than Bernard Shaw about correlation and causation.

To the list provided by Herne (1973), I would add the apparently innocuous manipulation of public opinion into a state of concern made in one of our UK national newspapers a few years ago with its headline: "*A half of all UK hospitals are below average.*"

Even before Huff (1954), the British Prime Minister Benjamin Disraeli (1804–1881) was attributed with the remark that there are "*lies, damned lies and statistics.*" So, by way of a guiding principle, one can reliably say that all of us are seekers after truth, but, as the Truth, Trust and Hope 2023 Symposium concluded, the future is education, and from that, we should always be continuing our learning and practicing our critical skills.

THE ROAD AHEAD

We can consider this at the levels of the individual, a scientific community (like the International Union of Crystallography), the national science academies, the federation of all scientists that is the International Science Council, and the leading organizations which we scientists interact with such as UNESCO and the United Nations. Then we must consider our interactions with governments as the proxy of the different countries' publics.

Individual scientists are rightly called upon more and more to explain our specific research. For myself an example would be my talks and writings on crystallography and sustainability which arose because of my IUCr community involvement, in turn based on my personal research expertize. In 2015, I summarized this as follows in the newsletter of the American Crystallographic Association following my talk on the subject at its 2015 Annual Conference (Helliwell 2015):

> Crystallography's traditional strengths yield the structures of molecules characterised as fully as X-rays, neutrons and/or electrons can via diffraction methods. Crystallography has a fine tradition in data archiving of coordinates and processed structure factor amplitudes, including introducing the crystallographic information framework, and now extending into raw diffraction images preservation. Our various methods and data are being applied in the characterisation for example of nanomaterials and led to our role within the CODATA/VAMAS project (https://codata.org/initiatives/previous-codata-working-groups/nanomaterials/) on proper "definition" of these new materials, so important for their precise safety descriptions. Crystallographers also take part in tackling some general issues for society at large including gender (in)equality in the careers of women versus men in science; we have much to be proud of as a field, but much remains to be done to avoid the current considerable loss of trained female scientists in future. Another aspect of sustainability is capacity building; prominent projects include the Middle East Synchrotron Radiation Source and the proposed African Synchrotron Light Source. These new facilities illustrate the importance of a sufficiently peaceful World and environment for research, pure and applied. In sustaining a peaceful World, prominent crystallographers such as Lawrence Bragg, Kathleen Lonsdale, Linus Pauling and Dorothy Hodgkin, and many others, made enormous contributions to World Peace, both within unavoidable war and against avoidable wars. In 2014 the International Year of Crystallography, with its numerous activities, successfully brought crystallography and its results to society at large and other research communities. The peaceful movement of scientists to Labs for training and research collaboration follow the International Science Council's free circulation of scientists statute. These are other aspects of sustaining our contributions to research and societal challenges.

The emphasis on the sustainability of our planet is a great concern, and because of misinformation and disinformation we must not only call out those untruths but also repeat our evidence on climate change, world health, and wildlife populations. A recent example is another USA National Academies of Science, Engineering, and Medicine with the Nobel Foundation 2021 conference "Our Planet, Our Future" (https://

www.nationalacademies.org/our-work/nobel-prize-summit-2021-our-planet-our-future). Also, very important is UNESCO's 1997 declaration on the "Responsibilities of the Present Generation to Future Generations" (https://en.unesco.org/about-us/legal-affairs/declaration-responsibilities-present-generations-towards-future-generations), which is a wonderful overarching guide for all of us in the scientific community, within which the declaration is:-

> The fate of future generations depends on decisions and actions taken today, and problems, including poverty, technological and material underdevelopment, unemployment, exclusion, discrimination and threats to the environment, must be solved in the interests of both present and future generations.

REFERENCES

Bragg, W. L. (1913) The structure of some crystals as indicated by their diffraction of X-rays. *Proc. R. Soc. London, Ser. A* 89, 248–277.

Bruno, I., Gražulis, S., Helliwell, J. R., Kabekkodu, S. N., McMahon, B. & Westbrook, J. (2017) Crystallography and databases. *Data Sci. J.* 16, 38. https://doi.org/10.5334/dsj-2017-038.

Fishkin, J. S. (2018) *Democracy When the People Are Thinking: Revitalizing Our Politics Through Public Deliberation*. Oxford University Press, Oxford.

Helliwell, J. R. (2015) *Crystallography and Sustainability*. Transactions of the American Crystallographic Association, 8–19. https://www.amercrystalassn.org/assets/volume45.pdf.

Helliwell, J. R. (2022) with colleagues from the IUCr committee on data and CODATA's CODATA international data policy committee and its subgroup on data rights and responsibilities. *Just Who Does Own Research Data?* https://forums.iucr.org/viewtopic.php?f=39&t=445.

Herne, H. (1973) Review of book How to lie with statistics by Huff, D. Pelican books. *J. R. Stat. Soc.* 22, 401–402.

Huff, D. (1954) *How to Lie with Statistics*. Pelican Books, London.

Leonelli, S. (2023) *Philosophy of Open Science* [Preprint]. http://philsci-archive.pitt.edu/id/eprint/21986 (accessed 2023-07-31).

Nobel Prize Foundation webinar on Truth, Trust and Hope, held May 2023 at the USA National Academies of Sciences, Engineering, and Medicine, Washington DC; recordings are available at https://www.nobelprize.org/events/nobel-prize-summit/2023. See also https://www.ipie.info/research/global-information-environment-trends-2023-an-expert-survey, https://digitalpublicgoods.net/information-pollution/ and https://10pointplan.org/wp-content/uploads/2023/05/10-Point-Plan-English.pdf.

UNESCO. (1997) Declaration on the "responsibilities of the present generation to future generations". https://en.unesco.org/about-us/legal-affairs/declaration-responsibilities-present-generations-towards-future-generations.

UNESCO recommendation on open science was adopted by the general conference of UNESCO at its 41st session, on 23 November 2021. https://en.unesco.org/science-sustainable-future/open-science/recommendation.

USA National Academies of Sciences, Engineering, and Medicine. (2018) *Open Science by Design: Realizing a Vision for 21st Century Research*. The National Academies Press, Washington, DC. https://nap.nationalacademies.org/catalog/25116/open-science-by-design-realizing-a-vision-for-21st-century.

USA National Academies of Science, Engineering and Medicine. (2019) *Reproducibility and Replicability in Science*. The National Academies Press, Washington, DC. https://nap.nationalacademies.org/catalog/25303/reproducibility-and-replicability-in-science

USA National Academies of Science, Engineering and Medicine with the Nobel Foundation. (2021) "Our planet, our future". https://www.nationalacademies.org/our-work/nobel-prize-summit-2021-our-planet-our-future.

Warren, J. E., Diakun, G., Bushnell-Wye, G., Fisher, S., Thalal, A., Helliwell, M. & Helliwell, J. R. (2008) Science experiments via telepresence at a synchrotron radiation source facility. *J. Synchrotron Rad.* 15, 191–194.

CHAPTER 10

Envoi

WHERE DOES MY SCIENTIFIC career fit into the themes of this book? I offer several of my scientific career objectives.

In "measuring the right thing," my training as a physicist is at the uppermost. This is in two ways. I made a dedicated effort to contribute to improving instrumentation and methods for understanding biological macromolecules. Central to this theme, I was very fortunate to join the UK's Synchrotron Radiation Source at Daresbury Laboratory, initially as a joint appointment with Keele University in 1979, where I was able to lead a team to develop the first dedicated protein crystallography beamline at the Daresbury SRS. This provided a step change in X-ray beam intensity and was fully tuneable, coming online in 1981. The Daresbury SRS was the world's first dedicated ("2nd generation") SR source. This led to my leading the European working group for macromolecular crystallography for the preparation of the world's first 3rd generation source, the European Synchrotron Radiation Facility (ESRF) (ESRF Foundation Phase Report 1987). The ESRF showed colossal increases in X-ray beam intensities from a new source, the X-ray undulator. The evolution of synchrotron radiation and the growth of its importance in crystallography I reviewed here (Helliwell 2012).

Within these synchrotron beamlines developments, I was determined to improve the detector provision and initiated collaborations with companies (Enraf Nonius, in the Netherlands, and Rigaku, in Japan), the UK's Rutherford Appleton Laboratory, and Daresbury Laboratory. This last arose due to the leadership of my senior colleague, Dr. Joan Bordas, later Director of the new Spanish Synchrotron Radiation Source "ALBA". Figure 10.1 (Top) shows the evolution of detector hardware for measuring the X-ray diffraction data, and Figure 10.1 (Bottom) shows what we were

DOI: 10.1201/9781003405399-10

After many, only partly successful, devices, the pixel detector finally combines all the desired electronic area detector properties for MX!

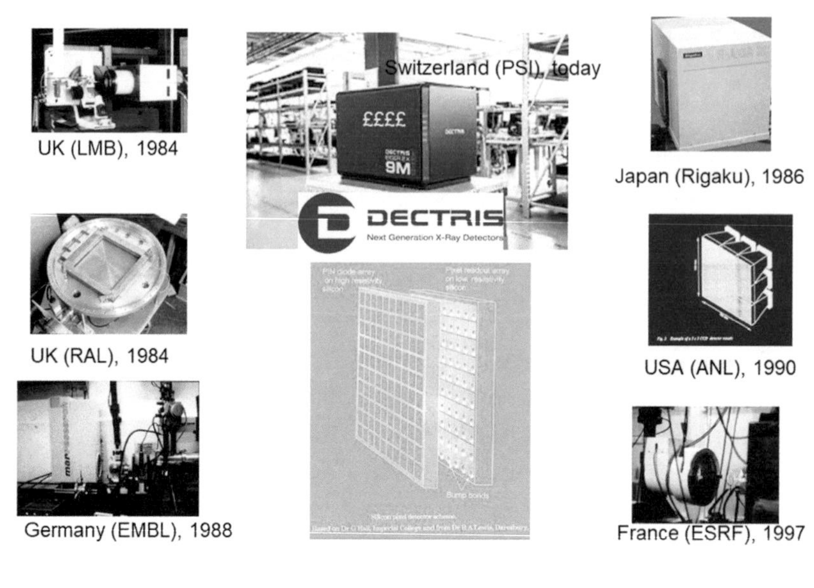

The detector challenge
The ideal detector
Photon counting detector (top) vs Image plate (bottom)
with weak signals: 10 & 0.3 secs exposure.
Extend the diffraction resolution via a better detector!

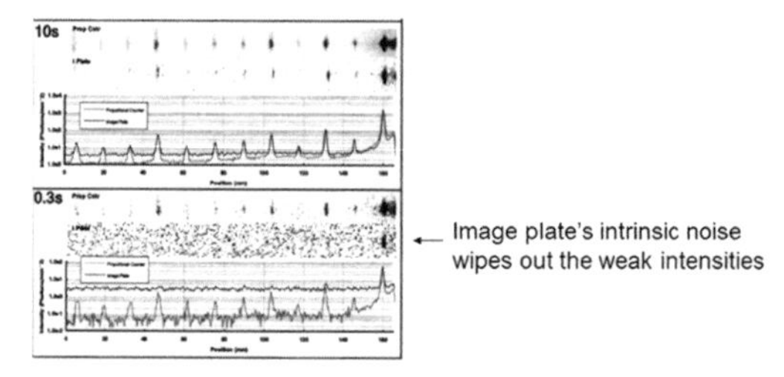

Rob Lewis, Daresbury Laboratory J. Synchrotron Rad. (1994). 1, 43-53

FIGURE 10.1 Top The evolution of detector hardware for measuring X-ray diffraction data. The pixel detector became the culmination of this evolution and was dominated by the excellent company Dectris in Switzerland with their range of excellent hardware. Bottom This work of my colleague of the time at the Synchrotron Radiation Source at Daresbury Laboratory, UK, Dr. Rob Lewis, showed up what we were striving for in terms of the ideal photon counting detector compared with the best integrating (analogue) detector of the time the image plate. Based on Helliwell (2022) lecture at CCP4 2022 for which a recording is available here: https://www.youtube.com/watch?v=I7e2EYrmBMg&list=PLrmG39_bWIGgvTugFLZaxHkR0WCkf8hQQ&index=24 .

striving for in terms of the ideal photon counting detector compared to the best integrating (analogue) detector of the time, the image plate.

These new synchrotron beamlines provided high intensity, variable wavelength, and variable bandpass opportunities, and their performances were enriched using the gradually improving detectors. Thereby, year on year, various benefits accrued in protein crystallography with synchrotron radiation, and spin-offs were possible into synchrotron chemical crystallography at the SRS and across into the neutron sources and facilities for neutron crystallography. Today, neutron protein crystallography is used, especially where previous studies with X-ray crystallography, NMR or electron microscopy on a system have failed to reveal an explanation of a protein's functional state and mechanism, such as an ionisable amino acid's protonation state or the orientation of bound water molecules in ligand molecular recognition binding sites. The molecular model of a protein determined by neutron diffraction will include experimentally determined deuterium positions (as hydrogen replacements) and have the advantages of being close to physiological temperature and undamaged by radiation. In neutron chemical crystallography, the neutron structure is regarded as the "true accuracy" (Sanjuan-Szklarz et al. 2020).

Naturally, after all the efforts of developing instrumentation and methods and after the new UK source being placed far away from Manchester, near Oxford, with the closure of the SRS in 2008, I needed a new career theme. This arose fortuitously when the International Union of Crystallography (IUCr) President of the time, Professor Dr. Sine Larsen, in 2011, asked me to lead their new IUCr working group, which tasked me to examine the benefits and costs of exploiting the new, large scale ("Big Data"), digital archives for preserving the experimental primary or raw diffraction data. We concluded in the affirmative that it was worthwhile to archive these data, certainly those that led to publications. The utility of unpublished "dark data" is still under debate but machine learning may increase their use. An interesting category of dark data is where the measuring team hit some sort of obstacle in fully analysing their raw data. At IUCr, we decided that this category warranted the launch of a new type of article at the IUCr's journal *IUCrData*, termed Raw Data Letters. These are described in Kroon Batenburg et al. (2022). These advertise data sets that are challenging in some way and thereby are an invitation from the measuring team to have others join in with the analyses.

These efforts continue with the IUCr's Committee on Data (https://www.iucr.org/resources/data/commdat), which I subsequently also chaired, until August 2023 after serving in these roles for 4 triennia, i.e., 12 years, the IUCr's maximum term of office. This is a good guideline to ensure new-blood volunteers can be appointed. This Big Data theme fits naturally under the motto of the Royal Society "Nullius in verba." To preserve the underpinning raw data is a wonderful opportunity for ensuring the reproducibility of all published scientific research. The trust of the public in science and its results can only increase with these impressive digital archiving developments. Each study then documents its "layers of truth" in a more comprehensive way. My last lecture as Chairman of the IUCr Committee on Data on this data theme I delivered at the American Crystallographic Association 2023 annual conference in Baltimore. I placed my slides and described the data session "*3.1.6 Validating models from the Data, Other Data and Theory*" at the ACA 2023 in Baltimore, USA at Zenodo: https://zenodo.org/record/8164848.

Are we measuring the right thing in terms of our biological macromolecules and their function in the living cell? I explored this theme first in writing up my talk that I had presented at the 2018 International Symposium of Diffraction Structural Biology held in Osaka, Japan (Nakagawa et al. 2021; Helliwell 2020). This is the toughest of all the challenges in my field of research. I suppose this is why I admire the research ongoing, although outside my own field, on batteries and their further development. They are a much, much simpler system than a living cell, but that system is of such great strategic importance to society at large and our planet, which is home to all of us irrespective of creed or race. Meanwhile, in my chosen research theme of biological macromolecules, I have over 100 Protein Data Bank depositions and publications on a variety of chemical themes: lectins and saccharide binding, enzymes and catalysis, marine crustacyanin with astaxanthin and colouration effects, and finally metallodrugs and metal imaging agents.

A major activity that I enjoy doing in my retirement is refereeing. These days, these are journal article submissions rather than research grant proposals. I summarized my efforts to ensure macromolecular crystallography articles are always assessed with their to be released Protein Data Bank files, as well as by the Protein Data Bank Validation Report, in what I entitled "Data science skills for referees: biological X-ray crystallography" (Helliwell 2018). As a detailed example of such a referee‘s report, I gave my permission for my referee's report (Helliwell 2018) to be publicly available to Nature Communications for the article by Langan et al. (2018).

REFERENCES

ESRF Foundation Phase Report (987) Grenoble, France.

Helliwell, J. R. (2012) The evolution of synchrotron radiation and the growth of its importance in crystallography. *Crystallogr. Rev.* 18(1), 33–93. https://doi.org/10.1080/0889311X.2011.631919.

Helliwell, J. R. (2018) *Referee's report on Langan et al (2018).* https://static-content.springer.com/esm/art%3A10.1038%2Fs41467-018-06957-w/MediaObjects/41467_2018_6957_MOESM1_ESM.pdf.

Helliwell, J. R. (2018) Data science skills for referees: Biological X-ray crystallography. *Crystallogr. Rev.* 24, 263–272.

Helliwell, J. R. (2020) What is the structural chemistry of the living organism at its temperature and pressure? *Acta Cryst. D* 76, 87–93.

Kroon-Batenburg, L. M. J., Helliwell, J. R. & Hester, J. R. (2022) IUCrData launches raw data letters. *IUCrData* 7, x220821.

Langan, P. S., Vandavasi, V. G., Weiss, K. L., Afonine, P. V., El Omari, K., Duman, R., Wagner, A. & Coates, L. (2018) Anomalous X-ray diffraction studies of ion transport in K+ channels. *Nat Commun* 9, 4540. https://doi.org/10.1038/s41467-018-06957-w

Nakagawa, A., Helliwell, J. R. & Yamagata, Y. (2021) Diffraction structural biology – An introductory overview. *Acta Cryst. D* 77, 278–279.

Sanjuan-Szklarz, W. F., Woińska, M., Domagała, S., Dominiak, P. M., Grabowsky, S., Jayatilaka, D., Gutmann, M. & Woźniak, K. (2020) On the accuracy and precision of X-ray and neutron diffraction results as a function of resolution and the electron density model. *IUCrJ* 7, 920–933.

Subject index

Name index